Integrating SOA and Web Services

RIVER PUBLISHERS SERIES IN INFORMATION SCIENCE AND TECHNOLOGY

Consulting Series Editor

KWANG-CHENG CHEN
National Taiwan University
Taiwan

Information science and technology enables 21st century into an Internet and multimedia era. Multimedia means the theory and application of filtering, coding, estimating, analyzing, detecting and recognizing, synthesizing, classifying, recording, and reproducing signals by digital and/or analog devices or techniques, while the scope of "signal" includes audio, video, speech, image, musical, multimedia, data/content, geophysical, sonar/radar, bio/medical, sensation, etc. Networking suggests transportation of such multimedia contents among nodes in communication and/or computer networks, to facilitate the ultimate Internet. Theory, technologies, protocols and standards, applications/ services, practice and implementation of wired/wireless networking are all within the scope of this series. We further extend the scope for 21st century life through the knowledge in robotics, machine learning, cognitive science, pattern recognition, quantum/biological/molecular computation and information processing, and applications to health and society advance.

- Communication/Computer Networking Technologies and Applications
- Queuing Theory, Optimization, Operation Research, Statistical Theory and Applications
- Multimedia/Speech/Video Processing, Theory and Applications of Signal Processing
- Computation and Information Processing, Machine Intelligence, Cognitive Science, and Decision

For a list of other books in this series, please visit www.riverpublishers.com.

Integrating SOA and Web Services

N. Sudha Bhuvaneswari

*School of Information Technology,
Dr. G.R. Damodaran College of Science,
Coimbatore, Tamilnadu, India*

and

S. Sujatha

*School of Information Technology,
Dr. G.R. Damodaran College of Science,
Coimbatore, Tamilnadu, India*

River Publishers

Routledge
Taylor & Francis Group

LONDON AND NEW YORK

Published 2011 by River Publishers
River Publishers
Alsbjergvej 10, 9260 Gistrup, Denmark
www.riverpublishers.com

Distributed exclusively by Routledge
4 Park Square, Milton Park, Abingdon, Oxon OX14 4RN
605 Third Avenue, New York, NY 10158

First published in paperback 2024

Integrating SOA and Web Services / by N. Sudha Bhuvaneswari, S. Sujatha.

Routledge is an imprint of the Taylor & Francis Group, an informa business

Publisher's Note
The publisher has gone to great lengths to ensure the quality of this reprint but points out that some imperfections in the original copies may be apparent.

While every effort is made to provide dependable information, the publisher, authors, and editors cannot be held responsible for any errors or omissions.

ISBN: 978-87-92329-65-3 (hbk)
ISBN: 978-87-7004-537-7 (pbk)
ISBN: 978-1-003-35718-6 (ebk)

DOI: 10.1201/9781003357186

Contents

Preface

Today's Information highway is developing faster and in a wider range as new technologies are added every day and hour. It is a real challenge for IT experts to stay abreast of all the developments in the industry. This can be attempted by reading up on these technological enhancements through books, journals and by surfing the net. Good books play a major role in helping to spread new developments. This book on Integrating SOA and Web Services has been designed based on the curriculum for Bachelors and Masters Degrees in Science, Engineering or Management and intends precisely to cover all the essential topics. Information Technology Professionals will also find the text useful as a reference source.

There are excellent books available today, which describe the internal workings of Service-Oriented Architecture and there are also separate books available for Web services. The purpose of this book is to integrate and to realize Service-Oriented Architecture with Web services which is one of the emerging technology in IT. The point is to link these two levels and demystify the subject. This book aims at doing this, in a step-by-step manner.

This book focuses on the latest technologies like Metadata Management, Security issues, Quality of Service and its commercialization. A chapter is also devoted to the study of Emerging standards and development tools for Enterprise Application Integration. Most enterprises have made extensive investments in system resources over the course of many years. Such enterprises have an enormous amount of data stored in legacy Enterprise Information Systems (EIS), so it is not practical to discard existing systems. It is more cost-effective to evolve and enhance EIS. But this could be done with the help of SOA realizing with Web services which is an emerging field in Information Technology.

SOA is usually realized through Web services. Web service specifications may add to the confusion of how to best utilize SOA to solve business problems. In order for a smooth transition to SOA, using an architectural style that helps in realizing Web services through SOA is necessary. Significant terms, major concepts and important statements are italicized. All the chapters include a summary of the chapter and ideas for review. An extensive index is added that provides rapid access to virtually any topic in the text by keyword. The book contains charts, diagrams, illustrations and each chapter lists extensive literature – books and papers are referenced. The book concentrates on this architecture, realization and integration of SOA with Web services and it consists of 12 chapters that are organized as follows:

Chapter 1 covers the goal for Integrating SOA with Web services as a world wide mesh of collaborating services, which are published and available for invocation on the Service Bus. It highlights the creation of a service-oriented environment and discusses the architecture of SOA, its technology perspective, policies, practices and frameworks to provide the right service at the right time.

Chapter 2 covers the importance of globalization and the nature of demanding customers who want agility and flexibility, and retailers. This chapter highlights the benefits offered by service orientation and the legacy system.

Chapter 3 covers governance and its role in any organization and an effective lifecycle approach. It discusses on proper deployment of effective lifecycle approach and it also covers the security and defense back-up and associated security rules imposed on SOA.

Chapter 4 covers the concepts and role of Business Process Management in Service-Oriented Architecture. It also deals with how to work with dynamic Business Process Management and SOA environment with discussion on the co-ordination between Business Process Management, Service-Oriented Architecture and Web Services with a short record of its progress.

Chapter 5 covers what a Web service is about and the different architectures and models that support efficient usage of Web based applications and Web services and it also discusses on the REST architecture and its implementation and working.

Chapter 6 covers ebXML electronic marketplace where enterprises of any size and in any geographical location can meet and conduct business with each other through the exchange of XML based messages. The specification, and evolution of ebXML is also discussed with emphasis on an open semantics framework.

Chapter 7 covers various concepts of integration of SOA and Web Services and the necessity and need for integration. This chapter also stresses the Web services semantics in WSDL and how carefully it must be designed before the actual implementation, and about the major roadblocks for Web services interoperability.

Chapter 8 covers metadata management and its role on building a Web service application. It elaborates on appropriate and potential technical challenges in maintaining the metadata. This chapter also summarizes the governance concept of SOA and what the future challenges are in the technological growth of SOA.

Chapter 9 covers the new set of requirements to the security landscape of SOA. This highlights how to implement security constraints to provide authentication, confidentiality and message integrity. It also discusses setting security at different levels like data level security, message level security. etc.

Chapter 10 covers ESB and its role as the infrastructure underpinning an integrated and flexible SOA. A clear definition of the interfaces, and of the capabilities and requirements of the services enables mediations to reconcile differences between service requestors and providers in ESB. We have discussed a range of mediation patterns and ESB usage patterns in which these abstract concepts are applied to enterprise scenarios.

Chapter 11 covers the functioning and working of Web Services Distribution Channel (WSDC) and how to create a WSDC component with its potentiality. Web Service Distribution Eco-System and the architecture that supports the WSDE environment and concludes with the SOA architecture and its role as middleware are treated.

Chapter 12 covers a summary of the emerging standards and trends in the integrated application of SOA and Web Services. It also highlights on the support provided by the Java technical suite to build an integrated system approach. A detailed description of WS-BPEL (Business Process Execution Language) is presented and the chapter concludes with a forecast on the future of EAI technologies.

Acknowledgements

We offer our heartfelt and sincere gratitude before the lotus feet of the Almighty for helping us to complete this book successfully. Thanks are also due to the following people who have helped make this first edition a reality, against all odds: The authors of this book sincerely thank Rivers Publishers, Denmark, for providing a sustainable atmosphere that has been a catalyst for us to compile this book. We received great inspiration and constant encouragement from the Management, Dr. G.R Damodaran College of Science, Coimbatore, India which we highly appreciate. We also sincerely thank our family members for their incredible support and understanding without which this text could never have come to fruition. We are very much grateful for the resources and authors of the resources that have been utilized for reference purposes in this book. We thank all the people who directly or indirectly rendered their support towards the successful completion of the book.

1

Introduction to SOA and Web Services

1.1 Evolution of SOA and Web Services

In the 1980s, applications were mostly vertical, built to meet the customer requirements in a vertical market segment. The software solutions were sufficient to meet the needs of a vertical industry. In the late 80s and early 90s, we saw the need for business applications to grow horizontally to cooperate with business partners. The industry saw the evolution of B2B (Business-to-Business) collaborations through components now spreading across several industry verticals. These components were distributed giving rise to an extended supply chain, providing customers and business partner's access to services. This is illustrated in Figure 1.1.

In today's world, the way that businesses operate has changed tremendously. Businesses not only want interaction with their partners, but they allow their customers and employees to access their business services electronically. In Business-to-Customer (B2C) customers have a direct access to the

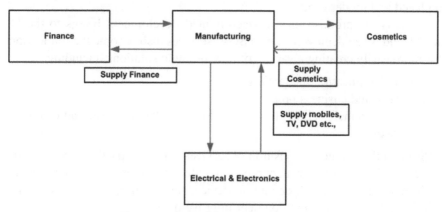

Figure 1.1 Extended supply chain scenario.

1

services offered by businesses. Exposing the business logic to an untrusted user base poses its own challenges in terms of security, integrity, and so on. Besides, such services must be user friendly and must hide the complexities of the internal business processes from the end customer.

This is where the true need for Service-Oriented Architecture is felt. Businesses should offer services rather than an interface to their business logic. The business logic is implemented in several components – exposing the interface to these components results in tight coupling with the business logic. A client application consumes the service through a well-defined interface to the service and does not care about how it is implemented.

If all applications were to use a common programming interface and interoperability protocol, the job of IT would be much simpler, complexity would be reduced, and existing functionality could be more easily reused. This is the promise that service-oriented development brings to the IT world, and when deployed using a Service-Oriented Architecture (SOA), services also become the foundation for more easily creating a variety of new strategic solutions.

Complexity is a fact of life in information technology (IT). Dealing with the complexity while building new applications, replacing existing applications, and keeping up with all the maintenance and enhancement requests represents a major challenge.

If all applications were to use a common programming interface and interoperability protocol, however, the job of IT would be much simpler, complexity would be reduced, and existing functionality could be more easily reused. After a common programming interface is in place, through which any application can be accessed, existing IT infrastructure can be more easily replaced and modernized.

This is the promise that service-oriented development brings to the IT world, and when deployed using a SOA, services also become the foundation for more easily creating a variety of new strategic solutions, including:

- Rapid application integration.
- Automated business processes.
- Multi-channel access to applications, including fixed and mobile devices.

A SOA facilitates the composition of services across disparate pieces of software, whether old or new; departmental, enterprise-wide, or inter-enterprise; mainframe, mid-tier, PC, or mobile device, to streamline IT processes and eliminate barriers to IT environment improvements.

These composite application solutions are within reach because of the widespread adoption of Web services and the transformational power of a SOA. The Web Services Description Language (WSDL) has become a standard programming interface to access any application, and SOAP has become a standard interoperability protocol to connect any application to any other. These two standards are a great beginning, and they are followed by many additional Web services specifications that define security, reliability, transactions, orchestration, and metadata management to meet additional requirements for enterprise features and qualities of service. Altogether, the Web services standards the best platform on which to build a SOA – the next-generation IT infrastructure.

1.2 Service-Oriented Enterprise

Driven by the convergence of key technologies and the universal adoption of Web services, the service-oriented enterprise promises to significantly improve corporate agility, speed time-to-market for new products and services, reduce IT costs, and improve operational efficiency.

As illustrated in Figure 1.2, several industry trends are converging to drive fundamental IT changes around the concepts and implementation of service orientation. The key technologies in this convergence are the following.

1.2.1 eXtensible Markup Language (XML)

eXtensible Markup Language (XML) is a common, independent data format across the enterprise and beyond that provides:

- Standard data types and structures, independent of any programming language, development environment, or software system.
- Pervasive technology for defining business documents and exchanging business information, including standard vocabularies for many industries.
- Ubiquitous software for handling operations on XML, including parsers, queries, and transformations.

1.2.2 Web Services

Web services are XML-based technologies for messaging, service description, discovery, and extended features, providing:

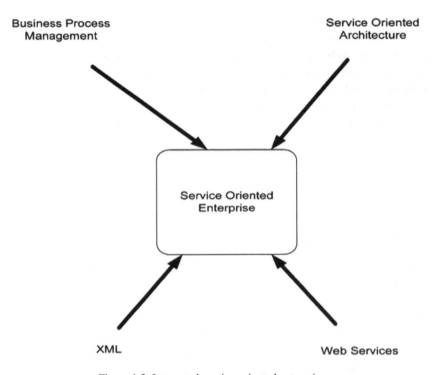

Figure 1.2 Integrated service-oriented enterprise.

- Pervasive, open standards for distributed computing interface descriptions and document exchange via messages.
- Independence from the underlying execution technology and application platforms.
- Extensibility for enterprise qualities of service such as security, reliability, and transactions.
- Support for composite applications such as business process flows, multi-channel access, and rapid integration.

1.2.3 Service-Oriented Architecture (SOA)

SOA presents a methodology for achieving application interoperability and reuse of IT assets that features:

- A strong architectural focus, including governance, processes, modeling, and tools.

- An ideal level of abstraction for aligning business needs and technical capabilities, and creating reusable, coarse-grain business functionality.
- A deployment infrastructure on which new applications can quickly and easily be built.
- A reusable library of services for common business and IT functions.

1.2.4 Business Process Management (BPM)

Business Process Management (BPM) are methodologies and technologies for automating business operations that:

- Explicitly describe business processes so that they are easier to understand, refine, and optimize.
- Make it easier to quickly modify business processes as business requirements change.
- Automate previously manual business processes and enforce business rules.
- Provide real-time information and analysis on business processes for decision makers.

1.2.5 Trends Converging to Create the Service-Oriented Enterprise

Individually, each of these technologies has had a profound effect on one or more aspects of business computing. When combined, they provide a comprehensive platform for obtaining the benefits of service orientation and taking the next step in the evolution of IT systems.

1.3 Service-Oriented Architecture (SOA)

A SOA is essentially a collection of services. These services communicate with each other. The communication can involve either simple data passing or it could involve two or more services coordinating some activity. Some means of connecting services to each other is needed.

SOAs are not a new thing. The first SOA for many people in the past was with the use DCOM or Object Request Brokers (ORBs) based on the CORBA specification.

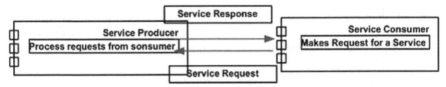

Figure 1.3 Web service interoperability.

1.3.1 Services

If a SOA is to be effective, we need a clear understanding of the term service. A service is a function that is well-defined, self-contained, and does not depend on the context or state of other services [2].

1.3.2 Connections

The technology of Web services is the most likely connection technology of SOAs. Web services essentially use XML to create a robust connection.

Figure 1.3 illustrates a basic SOA. It shows a service consumer at the right sending a service request message to a service provider at the left. The service provider returns a response message to the service consumer. The request and subsequent response connections are defined in some way that is understandable to both the service consumer and service provider. A service provider can also be a service consumer.

The term 'Web services' refers to the technologies that allow for making connections. Services are what you connect together using Web services. A service is the endpoint of a connection. Also, a service has some type of underlying computer system that supports the connection offered. The combination of services – internal and external to an organization – makes up a SOA.

It seems probable that eventually most software capabilities will be delivered and consumed as services. They may be implemented as tightly coupled systems, but the point of usage – to the portal, to the device, to another endpoint, and so on – will use a service-based interface. There is a comment that architects and designers need to be cautious to avoid everything becoming a service. It might be valid right now given the maturity of Web service protocols and technology to question whether everything is implemented using Web services, but that does not detract from the need to design everything from a service perspective. The service is the major construct for publishing and should be used at the point of each significant interface. SOA allows user to manage the usage (delivery, acquisition, consumption, and so on) in terms of, and in sets of, related services. This will have big implications for how we manage the software lifecycle – right from specification of requirements as services, design of services, acquisition and outsourcing as services, asset management of services, and so on.

Over time, the level of abstraction at which functionality is specified, published and or consumed has gradually become higher and higher. We have progressed from modules, to objects, to components, and now to services. However in many respects the naming of SOA is unfortunate. Whilst SOA is of course about architecture, it is impossible to constrain the discussion to architecture, because matters such as business design and the delivery process are also important considerations. A more useful nomenclature might be Service Orientation (or SO). There are actually a number of parallels with object orientation (or OO) and component-based development (CBD):

- Like objects and components, services represent natural building blocks that allow us to organize capabilities in ways that are familiar to us.
- Similarly to objects and components, a service is a fundamental building block that
  - Combines information and behavior.
  - Hides the internal workings from outside intrusion.
  - Presents a relatively simple interface to the rest of the organism.
- Where objects use abstract data types and data abstraction, services can provide a similar level of adaptability through aspect or context orientation.
- Where objects and components can be organized in class or service hierarchies with inherited behavior, services can be published and consumed singly or as hierarchies and or collaborations.

For many organizations, the logical starting place for investigating SOA is the consideration of Web services. However Web services are not inherently service oriented. A Web service merely exposes a capability that conforms to Web services protocols.

1.3.3 Principles and Definitions

The World Wide Web Consortium (W3C) for example refers to SOA as 'A set of components which can be invoked, and whose interface descriptions can be published and discovered'. It is a very technical perspective in which architecture is considered a technical implementation. This is odd, because the term architecture is more generally used to describe a style or set of practices – for example the style in which something is designed and constructed, for example Georgian buildings, Art Nouveau decoration or a garden by Sir Edwin Lutyens and Gertrude Jekyll.

CBDI believes a wider definition of SOA is required. In order to reach this definition, let us start with some existing definitions, and compare some W3C offerings with CBDI recommendations.

Service
A component capable of performing a task. A WSDL service: A collection of end points (W3C). A type of capability described using WSDL (CBDI).

A Service Definition
A vehicle by which a consumer's need or want is satisfied according to a negotiated contract (implied or explicit) which includes Service Agreement, Function Offered and so on (CBDI).

A Service Fulfillment
An instance of a capability execution (CBDI).

Web Service
- A software system designed to support interoperable machine-to-machine interaction over a network. It has an interface described in a format that machines can process (specifically WSDL). Other systems interact with the Web service in a manner prescribed by its description using SOAP messages, typically conveyed using HTTP with XML serialization in conjunction with other Web-related standards (W3C).
- A programmatic interface to a capability that is in conformance with WSnn protocols (CBDI) [2].

From these definitions, it will be clear that the W3C have adopted a somewhat narrower approach to defining services and other related artifacts than CBDI. CBDI differs slightly insofar as not all Services are Components, nor do they all perform a task. Also CBDI recommends it is useful to manage the type, definition and fulfillment as separate items. However it is in the definition of SOA that CBDI really parts company with the W3C.

SOA

- A set of components which can be invoked, and whose interface descriptions can be published and discovered (W3C).

CBDI rejects this definition on two counts: First the components (or implementations) will often not be a set. Second the W3C definition of architecture only considers the implemented and deployed components, rather than the science, art or practice of building the architecture. CBDI recommends SOA is more usefully defined as:

> The policies, practices, frameworks that enable application functionality to be provided and consumed as sets of services published at a granularity relevant to the service consumer. Services can be invoked, published and discovered, and are abstracted away from the implementation using a single, standards-based form of interface. (CBDI)

CBDI defines SOA as a style resulting from the use of particular policies, practices and frameworks that deliver services that conform to certain norms. Examples include certain granularity, independence from the implementation, and standards compliance. What these definitions highlight is that any form of service can be exposed with a Web services interface. However higher order qualities such as reusability and independence from implementation, will only be achieved by employing some science in a design and building process that is explicitly directed at incremental objectives beyond the basic interoperability enabled by use of Web services.

1.3.4 SOA Basics

Web services were merely a step along a much longer road. The notion of a service is an integral part of component thinking, and it is clear that distributed architectures were early attempts to implement SOA. What is important to recognize is that Web services are part of the wider picture that is SOA. The Web service is the programmatic interface to a capability that is in conformance with WSnn protocols. So Web services provide us with certain architectural characteristics and benefits – specifically platform independence, loose coupling, self description, and discovery – and they can enable a formal separation between the provider and consumer because of the formality of the interface.

> Service is the important concept. Web Services are the set of pro-
> tocols by which Services can be published, discovered and used in
> a technology neutral, standard form.

In fact, Web services are not a mandatory component of a SOA, although increasingly they will become so. SOA is potentially much wider in its scope than simply defining service implementation, addressing the quality of the service from the perspective of the provider and the consumer. COM and UML component packaging address components from the technology perspective, but CBD, or indeed Component-Based Software Engineering (CBSE), is the discipline by which you ensure you are building components that are aligned with the business. In the same way, Web services are purely the implementation. SOA is the approach, not just the service equivalent of a UML component packaging diagram.

Many of these SOA characteristics were illustrated in a recent CBDI report, which compared Web services published by two dotcom companies as alternatives to their normal browser-based access, enabling users to incorporate the functionality offered into their own applications. In one case it was immediately obvious that the Web services were meaningful business services – for example enabling the Service Consumer to retrieve prices, generate lists, or add an item to the shopping cart.

In contrast the other organization's services are quite different. It implemented a general purpose API, which simply provides Create, Read, Update, and Delete (CRUD) access to their database through Web services. While there is nothing at all wrong with this implementation, it requires that users understand the underlying model and comply with the business rules to ensure that their data integrity is protected. The WSDL tells nothing about the business or the entities. This is an example of Web services without SOA.

> SOA is not just an architecture of services seen from a technology
> perspective, but the policies, practices, and frameworks by which
> we ensure the right services are provided and consumed.

So, what we need is a framework for understanding what constitutes a good service and some Principles of Service Orientation that allow us to set policies and benchmarks.

We can discern two obvious sets here:

- Interface related principles: technology neutrality, standardization and consumability.

- Design principles: these are more about achieving quality services, meeting real business needs, and making services easy to use, inherently adaptable, and easy to manage.

Interestingly the second set might have been addressed to some extent by organizations that have established mature component architectures. However it is certainly our experience that most organizations have found this level of discipline hard to justify. While high quality components have been created perhaps for certain core applications where there is a clear case for widespread sharing and reuse, more generally it has been hard to incur what has been perceived as an investment cost with a short term return on investment.

However when the same principles are applied to services, there is now much greater awareness of the requirements, and frankly business and IT management have undergone a steep learning curve to better understand the cost and benefits of IT systems that are not designed for purpose. Here we have to be clear: not all services need all of these characteristics; however it is important that if a service is to be used by multiple consumers (as is typically the case when a SOA is required), the specification needs to be generalized, the service needs to be abstracted from the implementation (as in the earlier dotcom case study), and developers of consumer applications should not need to know about the underlying model and rules. The specification of obligations that client applications must meet needs to be formally defined and precise and the service must be offered at a relevant level of granularity that combines appropriate flexibility with ease of assembly into the business process.

Table 1.1 shows the principles of a good service design that are enabled by characteristics of either Web services or SOA.

If the principles summarized in Table 1.1 are complied with, we get some interesting benefits:

- *There is real synchronization between the business and IT implementation perspective.* For many years, business people have not really understood the IT architecture. With well designed services we can radically improve communications with the business, and indeed move beyond alignment and seriously consider convergence of business and IT processes.
- *A well-formed service provides us with a unit of management that relates to business usage.* Enforced separation of the service provision provides us with basis for understanding the lifecycle costs of a service and how it is used in the business.

Table 1.1 Web services and SOA.

Enabled by Web services	Technology neutral	Endpoint platform independence.
	Standardized	Standards-based protocols.
	Consumable	Enabling automated discovery and usage.
Enabled by SOA	Reusable	Use of Service, not reuse by copying of code/implementation.
	Abstracted	Service is abstracted from the implementation.
	Published	Precise, published specification functionality of service interface, not implementation.
	Formal	Formal contract between endpoints places obligations on provider and consumer.
	Relevant	Functionality presented at a granularity recognized by the user as a meaningful service.

- *When the service is abstracted from the implementation it is possible to consider various alternative options for delivery and collaboration models.* No one expects that, at any stage in the foreseeable future, core enterprise applications will be acquired purely by assembling services from multiple sources. However it is entirely realistic to assume that certain services will be acquired from external sources because it is more appropriate to acquire them. For example authentication services, a good example of third party commodity services that can deliver a superior service because of specialization, and the benefits of using a trusted external agency to improve authentication.

1.3.5 Process Matters

As indicated earlier, CBDI advises that good SOA is all about style – policy, practice and frameworks. This makes process matters an essential consideration.

Whilst some of the benefits of services might have been achieved by some organizations using components, there are relatively few organizations that rigorously enforce the separation of provision and consumption throughout the process. This gets easier with services because of the formality of the interface protocols, but we need to recognize that this separation needs managing. For example it is all too easy to separate the build processes of the service and the consumer, but if the consumer is being developed by the same team as the service then it is all too easy to test the services in a manner that reflects understanding of the underlying implementation.

With SOA it is critical to implement processes that ensure that there are at least two different and separate processes – for provider and consumer.

However, current user requirements for seamless end-to-end business processes, a key driver for using Web Services, mean that there will often be clear separation between the providing and consumer organizations, and potentially many to many relationships where each participant has different objectives but nevertheless all need to use the same service. Our recommendation is that development organizations behave like this, even when both the providing and consuming processes are in-house, to ensure they are properly designing services that accommodate future needs.

For the consumer, the process must be organized such that only the service interface matters, and there must be no dependence upon knowledge of the service implementation. If this can be achieved, considerable benefits of flexibility accrue because the service designers cannot make any assumptions about consumer behaviours. They have to provide formal specifications and contracts within the bounds of which consumers can use the service in whatever way they see fit. Consumer developers only need to know where the service is, what it does, how they can use it. The interface is really the only thing of consequence to the consumer as this defines how the service can be interacted with.

Similarly, whilst the provider has a very different set of concerns, it needs to develop and deliver a service that can be used by the Service Consumer in a completely separate process. The focus of attention for the provider is therefore again the interface – the description and the contract.

Another way of looking at this is to think about the nature of the collaboration between provider and consumer. At first sight you may think that there is a clear divide between implementation and provisioning, owned by the provider, and consumption, owned by the consumer. However if we look at these top level processes from the perspective of collaborations, then we see a very different picture.

What we have is a significant number of process areas where (depending on the nature of the service) there is deep collaboration between provider and consumer. Potentially we have a major reengineering of the software delivery process. Although we have two primary parties to the service-based process, we conclude there are three major process areas which we need to manage. Of course these decompose, but it seems to us that the following are the primary top level processes.

- *The process of delivering the service implementation.*
    - 'Traditional' development.
    - Programming.
    - Web services automated by tools.
- *The provisioning of the service – the lifecycle of the service as a reusable artefact.*
    - Commercial orientation.
    - Internal and external view.
    - Service level management.
- *The consumption process.*
    - Business process driven.
    - Service consumer could be internal or external.
    - Solution assembly from services, not code.
    - Increasingly graphical, declarative development approach.
    - Could be undertaken by business analyst or knowledge worker.

The advantage of taking this view is that the collaborative aspects of the process are primarily contained in the provisioning process area. And the provisioning area is incredibly important because the nature of the agreement has a major influence on the process requirements. There are perhaps two major patterns for designing consumer/provider collaborations:

- *Negotiated – Consumer and provider jointly agree service.* When new services are developed though, there is an opportunity for both provider and consumer to agree what and how the services should work. In industries where there are many participants all dealing with each other, and where services are common to many providers, it is essential that the industry considers standardizing those services. Examples include:
    - Early adopters.
    - New services.
    - Close partners.
    - Industry initiative – forming standards.
    - Internal use.
- *Instantiated – Take it or leave it.* One party in the collaborative scenario might simply dictate the services that must be used. Sometimes the service will already exist. You just choose to use it, or not. Examples include:
    - Dominant partner.

- Provider led – Use this service or we cannot do business.
- Consumer led – Provide this service or we cannot do business.
- Industry initiative – standards compliance.
- Existing system/interface.

1.3.6 Architectures

This process view that we have examined at is a prerequisite to thinking about the type of architecture required and the horizons of interest, responsibility and integrity. For SOA there are three important architectural perspectives as shown in Figure 1.1.

- *The Application Architecture.* This is the business facing solution which consumes services from one or more providers and integrates them into the business processes.
- *The Service Architecture.* This provides a bridge between the implementations and the consuming applications, creating a logical view of sets of services which are available for use, invoked by a common interface and management architecture.
- *The Component Architecture.* This describes the various environments supporting the implemented applications, the business objects and their implementations.

These architectures can be viewed from either the consumer or provider perspective. Key to the architecture is that the consumer of a service should not be interested in the implementation detail of the service – just the service provided. The implementation architecture could vary from provider to provider yet still deliver the same service. Similarly the provider should not be interested in the application that the service is consumed in. New unforeseen applications will reuse the same set of services.

The consumer is focused on their application architecture, the services used, but not the detail of the component architecture. They are interested at some level of detail in the general business objects that are of mutual interest, for example provider and consumer need to share a view of what an order is. But the consumer does not need to know how the order component and database are implemented.

Similarly, the provider is focused on the component architecture, the service architecture, but not on the application architecture. Again, they both need to understand certain information about the basic applications, for ex-

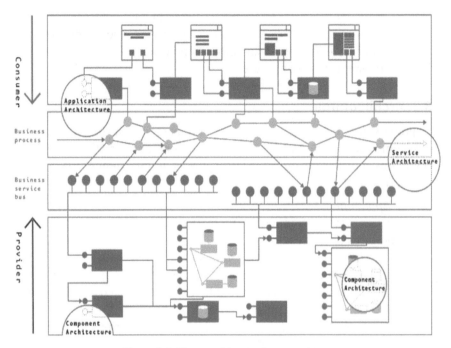

Figure 1.4 Three architectural perspectives.

ample to be able to set any sequencing rules and pre and post conditions. But the provider is not interested in every detail of the consuming application [3].

1.3.7 The Service Architecture

At the core of the SOA is the need to be able to manage services as first order deliverables. It is the service that we have constantly emphasized that is the key to communication between the provider and consumer. So we need a Service Architecture that ensures that services do not get reduced to the status of interfaces, rather they have an identity of their own, and can be managed individually and in sets.

CBDI developed the concept of the Business Service Bus (BSB) precisely to meet this need. The BSB is a logical view of the available and used services for a particular business domain, such as Human Resources or Logistics. It helps us answer questions such as:

- What service do I need?
- What services are available to me?

- What services will operate together (common semantics, business rules)?
- What substitute services are available?
- What are the dependencies between services and versions of services?

Rather than leaving developers to discover individual services and put them into context, the BSB is instead their starting point that guides them to a coherent set that has been assembled for their domain.

The purpose of the BSB is so that common specifications, policies, etc., can be made at the bus level, rather than for each individual service. For example, services on a bus should all follow the same semantic standards, adhere to the same security policy, and all point to the same global model of the domain. It also facilitates the implementation of a number of common, lower-level business infrastructure services that can be aggregated into other higher level business services on the same bus (for example, they could all use the same product code validation service). Each business domain develops a vocabulary and a business model of both process and object.

A key question for the Service Architecture is 'What is the scope of the service that is published to the Business Service Bus?' A simplistic answer is 'At a business level of abstraction'. However this answer is open to interpret-ation – better to have some heuristics that ensure that the service is the lowest common denominator that meets the criteria of business, and is consumer-oriented, agreed, and meaningful to the business. The key point here is that there is a process of aggregation and collaboration that should probably hap-pen separately from the implementing component as illustrated in Figure 1.2. By making it separate, there is a level of flexibility that allows the exposed service(s) to be adjusted without modifying the underlying components. In principle, the level of abstraction will be developed such that services are at a level that is relevant and appropriate to the consumer. The level might be one or all of the following:

- Business services.
- Service consumer-oriented.
- Agreed by both provider and consumer.
- Combine low-level implementation-based services into something meaningful to business.
- Coarser grained.
- Suitable for external use.
- Conforms to pre-existing connection design.

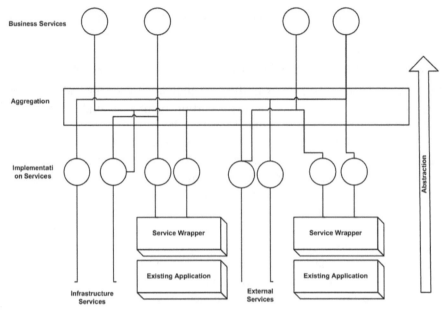

Figure 1.5 Levels of abstraction.

1.3.8 The SOA Platform

The key to separation is to define a virtual platform that is equally relevant to a number of real platforms. The objective of the virtual platform is to enable the separation of services from the implementation to be as complete as possible and allow components built on various implementation platforms to offer services which have no implementation dependency.

The virtual SOA platform comprises a blueprint which covers the development and implementation platforms. The blueprint provides guidance on the development and implementation of applications to ensure that the published services conform to the same set of structural principles that are relevant to the management and consumer view of the services.

When a number of different applications can all share the same structure, and where the relationships between the parts of the structure are the same, then we have what might be called a common architectural style. The style may be implemented in various ways; it might be a common technical environment, a set of policies, frameworks or practices. Example platform components of a virtual platform include:

- Host environment.

- Consumer environment.
- Middleware.
- Integration and assembly environment.
- Development environment.
- Asset management.
- Publishing & discovery.
- Service level management.
- Security infrastructure.
- Monitoring & measurement.
- Diagnostics & failure.
- Consumer/subscriber management.
- Web service protocols.
- Identity management.
- Certification.
- Deployment & Versioning.

1.3.9 The Enterprise SOA

The optimum implementation architecture for SOA is a component-based architecture. Many will be familiar with the concepts of process and entity component, and will understand the inherent stability and flexibility of this component architecture, which provide a one to one mapping between business entities and component implementations. Enterprise SOA (ESOA) brings the two main threads – Web services and CBD (or CBSE) – together. The result is an enterprise SOA that applies to both Web services made available externally and also to core business component services built or specified for internal use.

Example Enterprise Service-Oriented Architecture
The diagram in Figure 1.6 is an example of a SOA using Web services.

1.4 Understanding Web Services

1.4.1 Introducing a Web Service

Web services make software functionality available over the Internet so that programs like PHP, ASP, JSP, JavaBeans, the COM object, and all other favorite widgets can make a request to a program running on another server (a Web service) and use that program's response in a Website, WAP service, or other application. This can be done by passing parameter data with the request, for

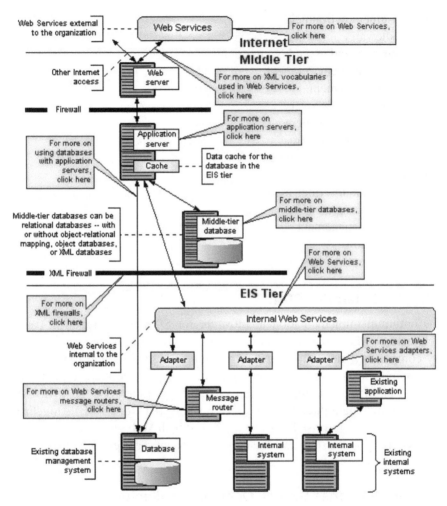

Figure 1.6 Enterprise Service-Oriented Architecture.

which a response could be received containing the result generated by the Web service.

The principles behind Web services are stunningly simple, and are nothing new in the world of distributed computing and the Internet:

- The Web service provider defines a format for requests for its service and the response the service will generate.
- A computer makes a request for the Web services across the network.

- The Web service performs some action, and sends the responses back. This action might be retrieving a stock quote, finding the best price for a particular product on the net, saving a new meeting to a calendar, translating a passage of text to another language, or validating a credit card number.

Standards Support

The sudden rise of interest in the services model is the incorporation of standard, open protocols for calling services and transmitting data. While in the past many data and service providers have had proprietary standards or rough-and-ready data formats, now users can rely on simple eXtensible Markup Language (XML) based access over plain old HTTP. This means easier access and should let developers working with all sorts of technologies start playing the Web services game.

The difference between Web services and technologies that developers have used in the past like DCOM, named pipes, and RMI, is that most Web services rely on open standards, are relatively easy to command, and have widespread support across the Unix/Windows divide.

The Simple Object Access Protocol (SOAP) is a W3C standard protocol that defines the format for Web service requests.

1.5 Integrating Web Services with SOA

SOA and Web services are two different things, but Web services are the preferred standards-based way to realize SOA. SOA is an architectural style for building software applications that use services available in a network such as the Web. It promotes loose coupling between software components so that they can be reused. Applications in SOA are built based on services. A service is an implementation of well-defined business functionality, and such services can then be consumed by clients in different applications or business processes.

SOA allows for the reuse of existing assets where new services can be created from an existing IT infrastructure of systems. In other words, it enables businesses to leverage existing investments by allowing them to reuse existing applications, and promises interoperability between heterogeneous applications and technologies. SOA provides a level of flexibility that was not possible before in the sense that:

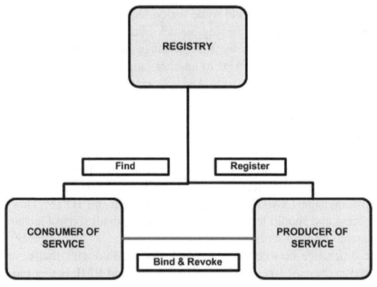

Figure 1.7 SOA's find-bind-execute paradigm.

- Services are software components with well-defined interfaces that are implementation-independent. An important aspect of SOA is the separation of the service interface (the what) from its implementation (the how). Such services are consumed by clients that are not concerned with how these services will execute their requests.
- Services are self-contained (perform predetermined tasks) and loosely coupled (for independence).
- Services can be dynamically discovered.
- Composite services can be built from aggregates of other services.

SOA uses the find-bind-execute paradigm as shown in Figure 1.7. In this paradigm, service providers register their service in a public registry. This registry is used by consumers to find services that match certain criteria. If the registry has such a service, it provides the consumer with a contract and an endpoint address for that service [1].

SOA-based applications are distributed multi-tier applications that have presentation, business logic, and persistence layers. Services are the building blocks of SOA applications. While any functionality can be made into a service, the challenge is to define a service interface that is at the right level of abstraction. Services should provide coarse-grained functionality.

1.5.1 Realizing SOA with Web Services

Web services are software systems designed to support interoperable machine-to-machine interaction over a network. This interoperability is gained through a set of XML-based open standards, such as WSDL, SOAP, and UDDI. These standards provide a common approach for defining, publishing, and using Web services.

Sun's Java Web Services Developer Pack 1.5 (Java WSDP 1.5) and Java 2 Platform, Enterprise Edition (J2EE) 1.4 can be used to develop state-of-the-art Web services to implement SOA. The J2EE 1.4 platform enables you to build and deploy Web services in your IT infrastructure on the application server platform. It provides the tools you need to quickly build, test, and deploy Web services and clients that interoperate with other Web services and clients running on Java-based or non-Java-based platforms. In addition, it enables businesses to expose their existing J2EE applications as Web services. Servlets and Enterprise JavaBeans components (EJBs) can be exposed as Web services that can be accessed by Java-based or non-Java-based Web service clients. J2EE applications can act as Web service clients themselves, and they can communicate with other Web services, regardless of how they are implemented.

1.6 Summary

The goal for integrating SOA with Web services is a worldwide mesh of collaborating services, which are published and available for invocation on the Service Bus. Adopting SOA is essential to deliver the business agility and IT flexibility promised by Web services. These benefits are delivered not by just viewing service architecture from a technology perspective and the adoption of Web service protocols, but require the creation of a Service-Oriented Environment that is based on the following key principals service is the important concept. Web services are the set of protocols by which services can be published, discovered and used in a technology neutral, standard form.

- SOA is not just architecture of services seen from a technology perspective, but the policies, practices, and frameworks by which we ensure the right services are provided and consumed.
- With SOA it is critical to implement processes that ensure that there are at least two different and separate processes – for provider and consumer.

- Rather than leaving developers to discover individual services and put them into context, the Business Service Bus is instead their starting point that guides them to a coherent set that has been assembled for their domain.

References

1. Abhishek Agrawal. Service-Oriented Architecture (SOA). Rightway Solution, http://www.rightwaysolution.com/soa.html, 2009.
2. Barry & Associates. Web Services and Service-Oriented Architecture. http://www.service-architecture.com/web-services/articles/service.html, 2010.
3. David Sprott and Lawrence Wilkes. Understanding Service Oriented Architecture. CBDI Forum, http://msdn.microsoft.com/en-us/library/aa480021.aspx, 2004.
4. Exforsys Inc. SOA Web Services – SOA Evolution. http://www.exforsys.com/tutorials/sws.html, 2010.

2

The Service Architecture

2.1 Introduction

A Service-Oriented Architecture (SOA) is the underlying structure support-ing communications between services. SOA defines how two computing entities, such as programs, interact in such a way as to enable one entity to perform a unit of work on behalf of another entity. Service interactions are defined using a description language. Each interaction is self-contained and loosely coupled, so that each interaction is independent of any other interaction.

Simple Object Access Protocol (SOAP)-based Web services are becom-ing the most common implementation of SOA. However, there are non-Web services implementations of SOA that provide similar benefits. The protocol independence of SOA means that different consumers can communicate with the service in different ways. Ideally, there should be a management layer between the providers and consumers to ensure complete flexibility regarding implementation protocols.

2.1.1 SOA Example Scenario

Whether you realize it or not, you have probably relied upon SOA, perhaps when you made a purchase online. Let us use Land's End as an example. You look at their catalog and choose a number of items. You specify your order through one service, which communicates with an inventory service to find out if the items you have requested are available in the sizes and colors that you want. Your order and shipping details are submitted to another service which calculates your total, tells you when your order should arrive and furnishes a tracking number that, through another service, will allow you to keep track of your order's status and location en route to your door. The entire process, from the initial order to its delivery, is managed by communications between the Web services programs talking to other programs, all made possible by the underlying framework that SOA provides.

2.2 SOA Services

It takes an end-to-end lifecycle approach to defining and implementing SOA solutions that meet a user's exact needs. Vendors work to clearly understand user needs and then determine the best possible approach with standard-based application and infrastructure environment, they help architect, consolidate, integrate, and improve it based on SOA. With the increased agility SOA Provides, user can find ways to align IT resources in support of business goals more easily and cost effectively.

SOA is the enabler for an agile IT and Business architecture and have thus become the corner stone of every company's survival and growth. If SOA solutions are designed to leverage on existing assets while creating a platform for enterprise transformation and optimization.

A wide range of SOA services that cover a variety of business applications includes:

- SOA Assessment and Envisioning.
- SOA Governance.
- SOA Enablement.
- SOA Service Development.
- SOA Delivery and Implementation.
- SOA Management.
- SOA Competency Centers.

2.3 Service Lifecycle

The objective of Service-Oriented Architecture (SOA) governance is to ensure that a SOA strategy delivers maximum value and SOA service lifecycle management is a fundamental SOA governance activity. The three processes within lifecycle management – service portfolio management, service consumption, and service creation – manage the planning, definition, development, and use of services. Tactically, service lifecycle management ensures that each service is of the highest possible quality and used appropriately. Strategically, service lifecycle management ensures that you are building the right portfolio of services to deliver the highest possible business value over time. Understanding the essentials of these processes and how they relate

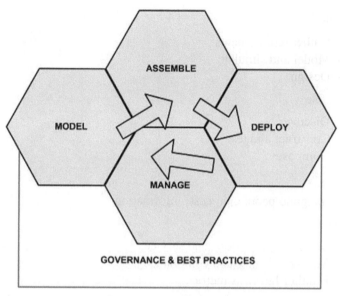

Figure 2.1 SOA lifecycle.

to each other enables architects to guide their firms' SOA strategy toward delivering consistent business value.

2.3.1 Phases of a SOA Lifecycle

We can distinguish the following phases of a SOA lifecycle:

- During the Model phase requirements are gathered and the desired business processes are modeled and specified.
- During the Assemble phase services are developed or assembled from existing services which need to be discovered from those approved services in the portfolio. The assemble phase also includes construction (or procurement) and testing of the resultant services.
- During the Deploy phase, the services are put into the operational environments and deliver the Business goals of integrating People, Processes and Information.
- During the Manage phase the applications and services are managed to ensure compliance with business policies and monitored to facilitate an understanding of how well the business or IT systems are performing and thus feedback into the overall business process improvement.

The SOA lifecycle can be summarized as follows:

- Model
    - Gather requirements
    - Model and simulate
    - Design
- Assemble
    - Discover
    - Construct and test
    - Compose
- Deploy
    - Integrate people, process, information
- Manage
    - Manage applications and services
    - Manage identity and compliance
    - Monitor business metrics
- Governance and best practices
    - Financial transparency
    - Business and IT alignment
    - Process control

2.3.2 Service Lifecycle Revisited

Services, like all other development assets and applications have their own lifecycle and as such need to be managed through their lifecycle state transitions. A service lifecycle generally models a typical SDLC with stages including design, development, test, QA, production, and deprecation. Many organizations will add versioning into the process between production and deprecation, although in reality each new version of a service will have its own lifecycle.

A SOA Governance product must be able to manage the lifecycle stage of a service and should provide a workflow-based process for migrating services between stages. Often this process will closely mirror the original publication process described above. It will include a set of policies that define criteria. A service must meet the needs before it can be migrated. It will also in many cases include manual approval steps.

The lifecycle stage of a service should be used to determine who can discover the service in the registry and who can access the service at runtime. It should also define which policy set is used to determine the run-time capabilities and requirements for accessing the service.

In the context of lifecycle management, the act of publishing a service to a registry so that it can be found by a broad audience of interested parties may seem like a simple enough task. In fact, this is one of the most basic, and yet most important functions of a SOA Governance solution [5].

The essence of governance can be easily captured in the phrase 'encouraging desired behavior'. This simple concept provides a backdrop to help understand what a governance solution should be focusing on, and the capabilities it should provide. Essentially it is not enough to merely provide a stick with which to beat developers and architects, the service must also provide a carrot to encourage people to participate in governance processes.

With this in mind, it is necessary to think about what is the desired behavior for the participants in a SOA. For many organizations, one of the most important aspects of SOA Governance is the process of ensuring that the services that are published are appropriate. 'Appropriate' in this context is another word a little like 'desired'. It can mean many things, but the reality is that an 'appropriate' service is a service that meets a set of criteria defined by the enterprise, often including the following:

- It is not a duplicate of, or similar to an existing service.
- It meets design criteria for transport, operation type, schema, etc.
- It is at an appropriate level of business functionality granularity (e.g. a 'top-down' design rather than 'bottoms-up').
- It is of broad interest and therefore likely to be reused.
- It complies with appropriate industry standards and recommendation (e.g. WS-I basic profile).

Some of these criteria can be readily automated like WS-I basic profile compliance, others will likely require manual steps. To this end, before a service can be published it should pass through a workflow process that will verify the automatable criteria before requiring a manual approval step. A well-designed SOA Governance solution will manage this workflow as a series of customizable, automatable defined process steps and will allow developers and approvers to see services at appropriate phases of this process.

SOA Software's Repository Manager and Policy Manager products combine to provide a comprehensive SOA Lifecycle Management solution. They share a common state-machine, and common metamodel providing seamless SOA asset lifecycle management capabilities [1].

Figure 2.2 Service deployment.

2.4 SOA Models

As the adoption rate of the Service-Oriented Architecture grows, it becomes more apparent that the level of abstraction, provided by Web Services APIs like JAX-RPC in Java or Web Services Extension (WSE) APIs in .Net, is not sufficient for effective SOA implementations.

The semantics of these APIs is geared more toward technical aspects of services invocations and SOAP processing, then towards service usage and support.

- The majority of them provide only SOAP over HTTP support, which is not always an optimal transport for SOA implementation.
- The majority of them provide only synchronous and one-way service invocation, which are only a subset of the service interaction styles.
- These APIs are directly exposed to the implementation code which leads to the following:
    - Business implementation code often gets intertwined with the service communications support, which makes its harder to implement, understand, maintain and debug.
    - Any API changes require changes in the business implementation.
- These APIs do not directly support many important service runtime patterns. For example, implementation of dynamic routing requires custom programming and usage of additional APIs (JAX-R in Java) for accessing registry.

In attempt to fix some of the issues of the current APIs sets there are currently attempts underway to raise the level of abstraction through defining SOA programming model which aim to reduce the complexity to which application developers are exposed to when they deal directly with middleware or Web services specific APIs. By removing the majority of communications support from the business code and hiding it behind programming model abstraction/implementation this approach facilitates:

- Simplified development of business services.
- Simplified assembly and deployment of business solutions built as networks of services.
- Increased agility and flexibility.
- Protection of business logic assets by shielding from low-level techno-logy change.
- Improved testability.

One of the first attempts to create such type of model was Web Services Invocation Framework (WSIF) originally introduced by IBM and currently part of Apache foundation.

WSIF attempted to align service usage model with the WSDL-based ser-vice definition-WSIF APIs directly support WSDL semantics. This enabled WSIF to provide a universal invocation model for different implementations of services over different transports. Although WSIF never gained wide ad-option by itself, it is used by many BPEL engines, for example WPC from IBM and BPEL Manager from Oracle, as an API for service invocations.

The three models currently gaining most popularity for SOA implement-ations are:

- Windows Communication Foundation (Indigo) Programming Model from Microsoft, which attempts to simplify service programming by creating a unified OO model for all service artifacts.
- Java Business Integration (JBI) model from Java Community Process, which address complexities and variabilities of services programming through creation of services abstraction layer in a form of a specialized (service) container.
- Service Components Architecture (SCA) from IBM, BEA, IONA, Oracle, SAP, Siebel, Sybase, etc., is based on the premise that a 'well-constructed component based architecture with well-defined in-terfaces and clear-cut component responsibilities can quite justifiable be considered as a SOA' [4].

These programming models attempt to go beyond just service invocations by seamlessly incorporating service orchestration support and many of the patterns required for successful SOA implementation. They also serve as a foundation for implementation of the Enterprise Service Bus.

Table 2.1 Relationship of SO entities to OO entities in Indigo (Source: Borris Lublinsky, 2006).

SO Entities	OO Entities	Annotations
Service contract	Interface	Annotate interface with [ServiceContract]
Service operation	Method	Annotate interface method with [Operation-Contract]
Implementation class	Class	Annotate class with [ServiceBehavior] and derive it from service contract interface
Implementation method	Method	Annotate method with [OperationBehavior]
Data contract	Class	Annotate class with [DataContract] and its members with [DataMember]

2.4.1 Indigo Programming Model

Indigo is the latest implementation of the programming model for service oriented architecture from Microsoft, supporting a rich set of technologies for 'creating, consuming, processing and transmitting messages'. Indigo is planed to be released with the next version of Windows Vista.

A service is defined in Indigo as a program exposing a collection of endpoints, with each endpoint defined as a combination of three major elements:

- The Endpoint's address-a network address through which the endpoint can be accessed.
- The Endpoint's Binding specifies additional detail defining communications with endpoint including transport protocol (e.g., TCP, HTTP) and policies-encoding (e.g., text, binary), security requirements (e.g., SSL, SOAP message security), etc.
- The Endpoint's Contract-specification of operations, exposed by this endpoint, messages, used by these operations and Message Exchange Patterns (MEPs) such as one-way, duplex, and request/reply.

Such definition effectively extends WSDL-based service definition by allowing exposing the same functionality (service) through multiple endpoints with different bindings and endpoint contracts (QoS). A foundation of Indigo's programming model is OO implementation of all aspects of SOA programming.

Table 2.1 shows the mapping (defined by Indigo programming model) between SO and OO concepts and the annotations that link them together. This fusion of SO and OO is simultaneously a major strength and weakness of Indigo:

- OO is a well-established paradigm familiar to the majority of developers. This means that programmers can start developing new Service-Oriented Architecture solutions using familiar techniques. Indigo runtime, in this case, converts, under the covers, OO style invocations into interoperable SOAP messages for communications.
- SO is significantly different from OO. Usage of OO as a mechanism for defining and implementing services can create a significant implementation mismatch (granularity, coupling, etc.), which might lead to suboptimal or even plainly bad SOA implementation.

Indigo supports two major approaches to the service invocation:

- RPC style invocations (both synchronous and asynchronous) carrying lists of typed parameters (initial Web Services vision). This type of service invocations is similar to the traditional methods invocations, used in distributed objects and RPC implementations, messaging style invocations (both synchronous and asynchronous). These types of service invocations are similar to the traditional messaging systems.
- Depending on the type of access service provides (RPC vs. Messaging), its contract is defined either in the form of interface (RPC) or message (Messaging) contract (see Table 2.1).

Another fundamental feature in Indigo is introduction of connectors – managed framework, providing secure and reliable communications – for accessing of service endpoints. Usage of connectors reduces the amount of the 'plumbing code' required for building of interoperable services, consequently simplifying creation of a web of 'connected systems'. This is achieved by separating 'plumbing' into separate classes (class hierarchies) and providing a 'standard' implementation in most cases.

The Indigo connectors use a small number of concepts (port, message, channel) to allow invoking of services independent of transport or target platform. The most important of these concepts is a channel, which represents a core abstraction for sending and receiving messages to and from a port. There are two categories of channels defined in Indigo:

1. Transport channels handle support for a specific transport, for example HTTP, TCP, UDP, or MSMQ and topologies, for example, point-to-point, end-to-end through intermediaries, peer-to-peer, publish and subscribe.
2. Protocol channels handle support for specific QoS characteristics, for example, security channel encrypts message and adds security header.

Indigo uses WS-*[c] specifications for implementation of protocol channels. Adherence to the standards makes Indigo's implementation interoperable with other systems, based on WS-* complaint implementation.

Indigo also supports the notion of channels composition-layering of channel on top of another channel. For example, a security protocol channel can be layered over HTTP transport channel to provide secure communications over HTTP.

The Indigo connector has a build in support for intermediaries including firewalls, proxies, and application-level gateways. These intermediaries are the foundation for implementation of many of the patterns required for successful SOA implementation, including message validation, security enforcement, transport switching, monitoring and management, load balancing and context-based routing.

In addition to supporting creation of the business services, Indigo provides several system services that can be used by any business service implementation. Examples of such services are:

- Identity federation. This service is based on WS-Federation and supports management and validation of identities both internally in the enterprise and from foreign trust boundaries. Its implementation provides authentication brokering between the service and the corresponding trust authority.
- Transactional support. This service is based on WSAtomicTransactions specification and simplifies service-based transactional programming (it supports SQL Server, ADO.NET, MSMQ, distributed transaction coordinator (DTC), etc).

2.4.2 JBI Programming Model

Capitalizing on the success of applications servers for hosting applications, Java community process has based its JBI implementation on a notion of a service container.

As defined in Java business integration, IEEE Internet Computing

> JBI is a pluggable architecture consisting of a container and plug-ins. The container hosts plug-in components that communicate via message routers. Architecturally, components interact via an abstract service model – a messaging model that resides at a level of abstraction above any particular protocol or message-encoding format.

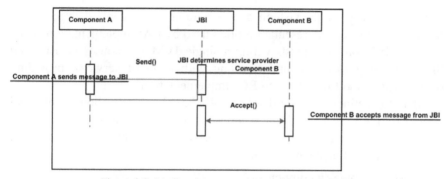

Figure 2.3 Mediated message exchange through JBI.

In the JBI-based implementation, services do not interact with each other directly. Instead, similar to the message broker architecture widely adopted in EAI implementations, JBI container acts as universal intermediary routing messages between services (Figure 2.3).

This separation of the participants in an exchange foundation of JBI architecture provides a layer of decoupling between service providers and consumers and simultaneously a well-defined place for mediating service traffic. Mediations, in this case, can support a wide variety of functionality, ranging from message transformation and security enforcement to content-based routing and service versioning.

JBI container hosts two distinct types of plug-in components:

- Service Engine (SE). SEs is essentially standard containers for hosting service providers and service consumers that are internal to the JBI environment. Example of SEs often present in JBI environment are data transformers, business rules containers and BPEL engines.
- Binding Component (BC). BCs provide connectivity to services consumers and providers external to a JBI environment. BCs allow integrating components/applications that do not provide Java APIs and use remote access technologies to access them.

Interactions between plug-in components employ standardized service definitions based on WSDL 2.0. WSDL 2.0 definitions provide shared understanding, between services consumers and providers and serves as a foundation of interoperability in JBI implementations.

In addition to standardized service definitions JBI uses the notion of 'normalized' messages supporting global components interoperability. Message normalization allows for mapping protocol and business specific context into a generic, transportable fashion and is similar in concept to introduction 'canonical' message representation often used by EAI implementations [2].

Each JBI container exists within a single JVM and houses all BCs and SEs, belonging to this container along with a set of services, providing operational support for SEs, and BCs implementation.

JBI also defines a standard set of JMX-based controls allowing external administrative tools to perform a variety of system administrative tasks, as well as administer the components themselves. This administration support provides standard mechanisms for

- Installing plug-in components.
- Managing plug-in component's lifecycle (stop/start, etc.)
- Deploying service artifacts to components.

2.4.3 Service Component Architecture

Although Service Component Architecture (SCA) is defined as a specification defining a model for building systems using Service-Oriented Architecture it is effectively a model for composing components into services extended to additionally support composition of services into solutions.

SOA is based on two main metamodels:

1. Type metamodel.
2. Composition metamodel.

2.4.3.1 Type Metamodel

This metamodel (Figure 2.4) describes component types, interfaces, and data structures.

A component implementation is defined by the following four groups of specifications:

- Provided interfaces-set of interfaces, defined by a component. These interfaces are usually defined as WSDL port type or language interface, for example Java or C++. A component can expose zero or more interfaces. Each interface is comprised by several methods.
- Required specifications (references)-set of interfaces used by a component's implementation. These interfaces are usually defined as WSDL port type or language interface, for example Java or C++. Component can have zero or more references.

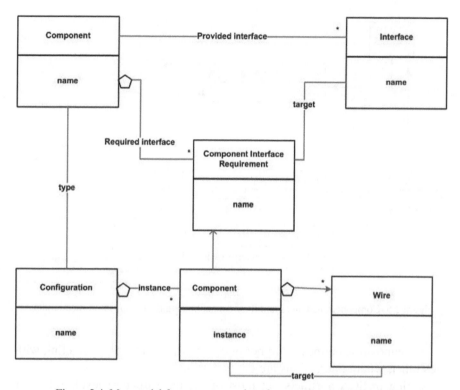

Figure 2.4 Metamodel for components, interfaces, and their dependencies.

- Properties that may be set on the component to tailor or customize its behavior. Each property is defined as a property element. Component can contain zero or more property elements.
- Implementation artifacts defining the implementation of the component. SCA allows for many different implementation technologies such as Java, BPEL, C++, SQL, etc SCA defines an extensibility mechanism for introducing new implementation types.

2.4.3.2 Composition Metamodel

This metamodel (Figure 2.5) defines component instances and how they are connected.

A notion of instance in this metamodel is different from the notion of instance used in OO. A component instance here is a component implementation with complete resolved set of properties, modifying component behavior for solving a particular problem.

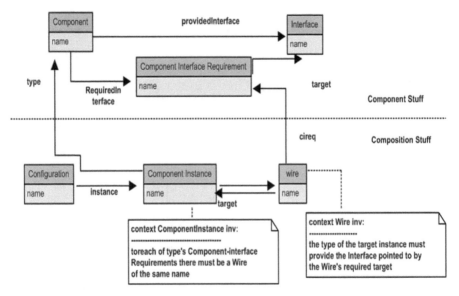

Figure 2.5 Component instances and their connections in the composition metamodel.

A composition defines a number of component instances, interacting with each other, where interactions are defined using wires.

Wiring in SCA abstracts out most of the infrastructure level code. For example, wires can be defined to be synchronous or asynchronous, support transactional behavior of components invocation, etc. SCA handles this infrastructure details under the covers. A wire can also connect between different interface languages (e.g. Java interfaces and WSDL portTypes/interfaces) in either direction, as long as the operations defined by the two interface types are equivalent.

In addition to wires SCA also supports module to module communications through special types of components-imports and exports. Combination of wiring, import and export components, allows for components referencing external services.

A component's composition, defined by composition metamodel is both similar and different from the service composition, although both of them define the way to make components/services work together. Service composition allows enhancing the functionality of participated services through functionality introduced by the composition itself, while this metamodel defines only connections between components.

SCA implementations rely on the Service Data Objects (SDO), providing the technology to represent a universal model for data. SDOs are the basis for the data exchange in component's interactions. The fundamental concept in the SDO architecture is the data object-a container holding primitive typed data and/or other data objects. Information about included data is provided by metadata, explicitly referenced by the data object. A combination of the data objects in SDO is represented by the data graph. In addition to the objects themselves, this graph contains a change summary that is used to log information about what data objects and properties in the graph have changed during processing. In addition to SDOs SCA introduces Service message objects (SMOs) which provides an abstraction layer for processing and manipulating messages exchanged between services (compare to normalized messages in JBI).

SCA is currently in its infancy and does not support the majority of patterns required for SOA implementation. Instead, current SCA implementation from IBM-Websphere ESB/WPS 6.0 introduces mediation framework, based on SCA and providing a well-defined mechanisms for mediation implementation and placing.

SCA implementation is especially powerful, when it is supported by GUI, allowing for graphical connection of the components on the palette, the way it is implemented in IBM's WebSphere Integration developer [2].

2.5 Principles of SOA

The core principles of Service Oriented Architecture focuses on service abstraction, service reusability, service composability, service autonomy, service optimization, service discoverability, service-orientation and interoperability, standardized service contract.

2.5.1 Explicit Boundaries

Everything needed by the service to provide its functionality should be passed to it when it is invoked. All access to the service should be via its publicly exposed interface, no hidden assumptions must be needed to invoke the service. 'Services are inextricably tied to messaging in that the only way into and out of a service are through messages'. A service invocation should as a general pattern not rely on a shared context, instead service invocations should be modeled as stateless. An interface exposed by a service is governed by a contract that describes its functional and non-functional capabilities and characteristics. The invocation of a service is an action that has a business effect, is possibly expensive in terms of resource consumption, and introduces a category of errors different than those of a local method invocation or remote procedure call. A service invocation is not a remote procedure call.

While consuming and providing services it should be certainly as easy as possible, it is therefore undesirable to hide too much of the fact that an interaction with a service takes place. The message sent to or received from the service, the service contract, and the service itself should all be first-class constructs within the SOA. This means, for example, that programming models and tools that are used should at least provide an API that exposes these concepts to the service programmer. In summary, a service exposes its functionality through an explicit interface that encapsulates its internals, interaction with a service is an explicit act, relying on the passing of messages between consumer and provider.

2.5.2 Shared Contract and Schema, Not Class

Starting from a service description (a contract), both a service consumer and a service provider should have everything they need to consume or provide the service. Following the principle of loose coupling, a service provider cannot rely on the consumer's ability to reuse any code that it provides in its own environment; after all, it might be using a different development or runtime environment. This principle puts severe limits on the type of data that can be exchanged in a SOA. Ideally, the data is exchanged as XML documents that could be validated against one or more schemas, since these are supported in every programming environment one can imagine. As a consequence, adherence to this principle is not possible in DCOM-based or RMI-based environments – which basically rule them out as a valid option for SOA.

2.5.3 Policy-Driven

To interact with a service, two orthogonal requirement sets have to be met:

1. The functionality, syntax and semantics of the provider must fit the consumer's requirements.
2. The technical capabilities and needs must match.

For example, a service provider may offer exactly the service a consumer needs, but offer it over JMS while the consumer can only use HTTP (e.g. because it is implemented on the .NET platform); a provider might require message-level encryption via the XML Encryption standard, while the consumer can only support transport-level security using SSL. Even in those cases where both partners do have the necessary capabilities, they might need to be 'activated' – e.g. a provider might encrypt response messages to different consumers using different algorithms, based on their needs.

To support access to a service from the largest possible number of differently equipped and capable consumers, a policy mechanism has been introduced as part of the SOA tool set. While the functional aspects are described in the service interface, the orthogonal, non-functional capabilities and needs are specified using policies.

2.5.4 Autonomous

Related to the explicit boundaries principle, a service is autonomous, in that its only relation to the outside world and from the SOA perspective – is through its interface. In particular, it must be possible to change a service's runtime environment, e.g. from a lightweight prototype implementation to a full-blown, application server-based collection of collaborating components, without any effect on its consumers. Services can be changed and deployed, versioned and managed independently of each other. A service provider cannot rely on the ability of its consumers to quickly adapt to a new version of the service, some of them might not even be able, or willing, to adapt to a new version of a service interface at all.

2.5.5 Service as Wire Formats, Not Programming Language APIs

Services are exposed using a specific wire format that needs to be supported. This principle is strongly related to the first two principles, but introduces a new perspective: To ensure the utmost accessibility and long-term usability, a service must be accessible from any platform that supports the exchange of messages adhering to the service interface as long as the interaction conforms to the policy defined for the service. For example, it is a useful test for conformance to this principle to consider whether it is possible to consume or provide a specific service from a mainstream dynamic programming language such as Perl, Python or Ruby. Even though none of these may currently play any role in the current technology landscape, this consideration can serve as a litmus test to assess whether the following criteria are met:

- All message formats are described using an open standard, or a human readable description.

- It is possible to create messages adhering to those schemas with reasonable effort without requiring a specific programmer's library.
- The semantics and syntax for additional information necessary for successful communication, such as headers for purposes such as security or reliability, follow a public specification or standard.
- At least one of the transport (or transfer) protocols used to interact with the service is (or is accessible via) a standard network protocol.

2.5.6 Document Orientation

To interact with services, data is passed as documents. A document is an explicitly modeled, hierarchical container for data. Self-descriptiveness is one important aspect of document-orientation. Ideally, a document will be modeled after real-world documents, such as purchase orders, invoices, or account statements. Documents should be designed so that they are useful on the context of a problem domain, which may suggest their use with one or more services.

Similarly to a real-world paper document, a document exchanged with a service will include redundant information. For example, a customer ID might be included along with the customer's address information. This redundancy is explicitly accepted since it serves to isolate the service interface from the underlying data model of both service consumer and service provider. When a document-oriented pattern is applied, service invocations become meaningful exchanges of business messages instead of context-free RPC calls. While not an absolute required, it can usually be assumed that XML will be used as the document format/syntax.

Messages flowing between participants in a SOA connect disparate systems that evolve independently of each other. The loose coupling principle mandates that the dependence on common knowledge ought to be as small as possible. When messages are sent in a Distributed Objects or RPC infrastructure, client and server can rely on a set of proxy classes (stubs and skeletons) generated from the same interface description document. If this is not the case, communication ceases on the assumption that the contract does not support interaction between those two parties. For this reason, RPC-style infrastructures require synchronized evolution of client and server program code. This is illustrated by the following comparison. Consider the following message:

and compare it to

$$2010\text{-}10\text{-}13$$
$$4911$$
$$3$$

While it is obvious that the second alternative is human-readable while the first one is not, it is also notable that in the second case, a participant that accesses the information via a technology such as XPath will be much better isolated against smaller, non-breaking changes than one that relies on the fixed syntax. Conversely, using a self-descriptive message format such as XML while still using RPC patterns, such as stub and skeleton generation, serves only to increase XML's reputation as the most effective way to waste bandwidth. If one uses XML, the benefits should be exploited, too.

2.5.7 Loosely Coupled

Most SOA proponents will agree that loose coupling is an important concept. Unfortunately, there are many different opinions about the characteristics that make a system 'loosely coupled'. There are multiple dimensions in which a system can be loosely or tightly coupled, and depending on the requirements and context, it may be loosely coupled in some of them and tightly coupled in others. Dimensions include:

- *Time*: When participants are loosely coupled in time, they do not have to be up and running at the same time to communicate. This requires some way of buffering/queuing in between them, although the approach taken for this is irrelevant. When one participant sends a message to the other one, it does not rely on an immediate answer message to continue processing.
- *Location*: If participants query for the address of participants they intend to communicate with, the location can change without having to re-program, reconfigure or even restart the communication partners. This implies some sort of lookup process using a directory or address that stores service endpoint addresses.
- *Type*: In an analogy to the concept of static vs. dynamic and weak vs. strong typing in programming languages, a participant can either rely on all or only on parts of a document structure to perform its work.
- *Version*: Participants can depend on a specific version of a service interface, or be resilient to change. The more exact the version match has to be, the less loosely coupled the participants. A good principle

to follow is Postel's Law: Service providers should be implemented to accept as many different versions as possible, and thus be liberal in what they accept, while service consumers should do their best to conform to exact grammars and document types. This increases the overall system's stability and flexibility.

- *Cardinality*: There may be a 1:1-relationship between service consumers and service providers, especially in cases where a request/response interaction takes place or an explicit message queue is used. In other cases, a service consumer which in this case is more reasonably called a 'message sender' or 'event source' may neither know nor care about the number of recipients of a message.
- *Lookup*: A participant that intends to invoke a service can either rely on a physical or logical name of a service provider to communicate with, or it can perform a lookup operation first, using a description of a set of capabilities instead. This implies a registry and/or repository that is able to match the consumer's needs to providers capabilities.
- *Interface*: Participants may require adherence to a service-specific interface or they may support a generic interface. If a generic interface is used, all participants consuming this generic interface can interact with all participants providing it. The principle of a single generic interface is at the core of the WWW's architecture.

It is not always feasible or even desirable to create a system that is loosely coupled in all of the dimensions mentioned above. For different types of services, different trade-offs need to be made.

2.5.8 Standards Compliance

A key principle to be followed in a SOA approach is the reliance on standards instead of proprietary APIs and formats. Standards exist for technical aspects such as data formats, metadata, transport and transfer protocols, as well as for business-level artifacts such as document types.

While this may seem absolutely obvious to many, some argue that a proprietary solution, such as those provided by some EAI or messaging vendors, follows SOA principles. This principle highlights the importance of standards – the more, the better. The most important aspect of any standard is its acceptance.

2.5.9 Vendor Independence

No architectural principle should rely on any particular vendor's product. To transform an abstract concept into a concrete, running system, it is unavoidable to decide on specific products, both commercial and free/open source software. None of these decisions must have implications on an architectural level. This implies reliance on both interoperability and portability standards as much as reasonably possible. As a result, a service provider or service consumer can be built using any technology that supports the appropriate standards, not restricted by any vendor roadmap.

2.5.10 Metadata-Driven

All of the metadata artifacts within the overall SOA need to be stored in a way that enables them to be discovered, retrieved and interpreted at both design and run time. Artifacts include descriptions of service interfaces, participants, endpoint and binding information, organizational units and responsibility, document types/schemas, consumer/provider relationships etc. As much as possible, usage of these artifacts should be automated by either code generation or interpretation and become part of the service and participant lifecycle.

2.6 SOA Mapping Components

Abstracting application functionality in the service provider Web services essentially provides a service wrapper that abstracts functionality across one or more applications. Some Web services may provide functionality from a single application while others provide an aggregate interface for multiple applications. The mapping of applications to services allows applications and technology to align with the business processes. Figure 2.6 shows an example of how application functionality is mapped to a service implementation.

Component architectures, like Enterprise Java Beans (EJBs), provide an encapsulation of services that can be delivered in an application server environment. A Web server provides the HTTP network transport for accessing the service. The application server hosts the SOAP interface and the object components that make up the service. The object components provide the business service layer above the applications.

The end result is that the Web service abstracts underlying applications to provide richly defined services. These services then map into well-defined business processes. Creating an abstraction layer above applications enables

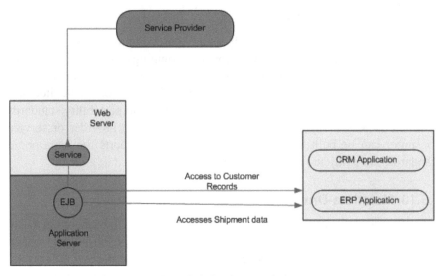

Figure 2.6 Service mapping.

them to deliver functionality in a more meaningful way defined as business services.

2.6.1 Service Contracts with WSDL

The service provider publishes its interface contract in the form of a Web Services Description Language (WSDL) document. The WSDL document defines the characteristics of the service including the location of the service and how to access it. From a development perspective, WSDL provides the contract that both the service requestor and service provider agree upon.

Once you define the business processes and organize services that can implement those processes, the next step is defining services and specific methods within the services. The WSDL document is an XML format of the specifications for the services. Within the WSDL document, you specify how to access the service, what methods it has available, how to access the methods, and what parameters to pass to the methods, and so on.

Service requestors locate services by using a service registry. In a Web services implementation, the service registry is implemented using Universal Description, Discovery, and Integration (UDDI). The UDDI server contains the database of service descriptions and provides them to the service requestor application. The requestor uses the WSDL document to understand the interface contract with the Web Service.

In general, WSDL documents are not physically stored in the UDDI server; rather the UDDI server provides URLs which identify to the service requestor where to find the WSDL document.

In programmer terms, the WSDL document defines the abstract interface used to access the service. From a design perspective, the WSDL must come before the service requestor and the service provider since it defines the contract between them. Once the WSDL has been designed and agreed upon, the service provider developers and the service requestor developers can work somewhat in parallel.

2.6.2 Service Discovery

Services are discovered by querying the UDDI server and accessing the WSDL documents for the services. By using a central registry, client applications can dynamically discover and bind to services. The UDDI server and protocol is reasonably generic. For a service architecture, the UDDI component provides a central service agency that knows about all the services within its domain. Whenever a service requestor needs to access or locate a service, the UDDI server is queried for information. UDDI provides a variety of ways to search the registry [3].

Once an organization and service have been located, the service requestor extracts the WSDL location. The WSDL is usually located with the service itself. Next, the requestor accesses the WSDL document to understand the specifics of the service. Using the information in the WSDL document, the requestor will understand how to access the service, what methods it has, and what parameters need to be sent, among other things. Figure 2.7 illustrates how this process works.

2.7 Summary

In this changing business climate, where globalization and demanding customers require agility and flexibility, retailers' IT systems are not helping them in meeting these expectations. Most of the legacy retail systems were built following tightly coupled point-to-point integration principles. Service orientation offers great benefits in moving from these legacy systems to more flexible and agile systems. This Service Architecture looks at the challenges

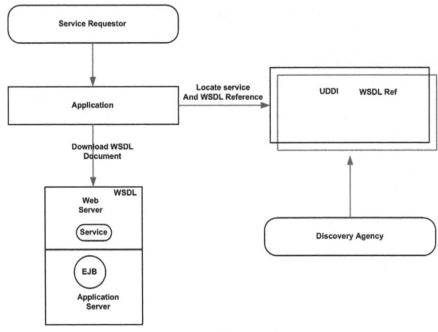

Figure 2.7 Process of service discovery.

faced by retailers, what is service orientation, and how it can benefit the retail enterprises.

References

1. Barry & Associates. Web Services and Service-Oriented Architecture. http://www.service-architecture.com/web-services/articles/service.html, 2010.
2. Borris Lublinsky. SOA Programming Models. InfoQ, http://www.infoq.com/articles/SOA-programming-models, 2006.
3. Krzysztof Brzostowski, Witold Rekuć, Janusz Sobecki, and Leopold Szczurowski. *Service Discovery in the SOA System*, Lecture Notes in Computer Science, Vol. 5991. Springer, 2010.
4. Michael Beisiegel et al. Service Component Architecture. Building Systems Using a Service Oriented Architecture. A Joint Whitepaper by BEA, IBM, Interface21, IONA, Oracle, SAP, Siebel, Sybase, 2005.
5. IBM. WSRR and the SOA Service Life Cycle. http://publib.boulder.ibm.com/infocenter/sr/v6r0/index.jsp?topic=/com.ibm.sr.doc/cwsr_overview_overview06.html, 2010.
6. SOA Software. Service Lifecycle Management. http://www.soa.com/solutions/service_lifecycle_management/, 2009.

3

Essence of SOA Governance

3.1 Governance

Governance is the activity of governing. It relates to decisions that define expectations, grant power, or verify performance. It consists either of a separate process or of a specific part of management or leadership processes. The word governance derives from the Greek verb 'kubernáo' which means to steer and was used for the first time in a metaphorical sense by Plato. It then passed on to Latin and then on to many languages. In general terms, governance occurs in three broad ways:

1. Through networks involving public-private partnerships (PPP) or with the collaboration of community organizations.
2. Through the use of market mechanisms whereby market principles of competition serve to allocate resources while operating under government regulation.
3. Through top-down methods that primarily involve governments and the state bureaucracy.

3.1.1 Types of Governance

3.1.1.1 Corporate Governance

Corporate governance consists of the set of processes, customs, policies, laws and institutions affecting the way people direct administer or control a corporation. Corporate governance also includes the relationships among the many players involved (the stakeholders) and the corporate goals. The principal players include the shareholders, management, and the board of directors. Other stakeholders include employees, suppliers, customers, banks and other lenders, regulators, the environment and the community at large.

3.1.1.2 Project Governance

The term governance as used in industry (especially in the information technology (IT) sector) describes the processes that need to exist for a successful project.

3.1.1.3 Information Technology Governance

IT Governance primarily deals with connections between business focus and IT management. The goal of clear governance is to assure the investment in IT general business value and mitigate the risks that are associated with IT projects.

The other types of governance include Participatory Governance, Non-Profit Governance, Measuring Governance.

3.2 Overview of SOA Governance

Service-Oriented Architecture (SOA) governance is a concept used for activities related to exercising control over services in a SOA. SOA governance can be seen as a subset of IT governance which itself is a subset of Corporate governance. The focus is on those resources to be leveraged for SOA to deliver value to the business. SOA requires a number of IT support processes as well as organizational processes that will also involve the business leaders. SOA needs a solid foundation that is based on standards and includes policies, contracts and service level agreements. The definitions of SOA governance agree in its purpose of exercising control, but differ in the responsibilities it should have. Some narrow definitions focus on imposing policies and monitoring services, while other definitions use a broader business-oriented perspective.

3.2.1 Scope of SOA Governance

Some typical governance issues that are likely to emerge in a SOA are:

- Delivering value to the stakeholders: investments are expected to return a benefit to the stakeholders – this is equally true for SOA.
- Compliance to standards or laws: IT systems require auditing to prove their compliance to regulations like the Sarbanes-Oxley Act (SOX). In a SOA, service behavior is often unknown.
- Change management: changing a service often has unforeseen consequences as the service consumers are unknown to the service pro-

viders. This makes an impact analysis for changing a service more difficult than usual.

- Ensuring quality of services: The flexibility of SOA to add new services requires extra attention for the quality of these services. This concerns both the quality of design and the quality of service. As services often call upon other services, one malfunctioning service can cause damage in many applications.

Some key activities that are often mentioned as being part of SOA governance are:

- Managing the portfolio of services: planning development of new services and updating current services.
- Managing the service lifecycle: meant to ensure that updates of services do not disturb current service consumers.
- Using policies to restrict behavior: rules can be created that all services need to apply to, to ensure consistency of services.
- Monitoring performance of services: because of service composition, the consequences of service downtime or underperformance can be severe. By monitoring service performance and availability, action can be taken instantly when a problem occurs.

3.2.2 Necessity of SOA Governance

SOA introduces many independent and self-contained moving parts or components which are reused widely across the enterprise and are a vital part of mission-critical business processes. SOA governance is about managing the quality, consistency, predictability, change and interdependencies of services. It is about blending the flexibility of service orientation with the control of traditional IT architectures [1].

3.3 Organization of SOA Governance

3.3.1 Architecture

Architectural policies provide the foundation and framework for SOA and enable user to build it better, faster, and cheaper. Every system must be built so that it both fits into the existing environment and reflects the organization's future vision and SOA strategy. Building out SOA to enable change is best done using an architectural approach that sets up a minimal set of constraints, thereby realizing consistency in service implementation, improved interoperability, stakeholder innovation, and enablement of applications that are minimally developed, yet offer general-purpose capabilities that are useful to other applications and take advantage of and enhance a shared infrastructure.

3.3.2 Technology Infrastructure

Technology must be identified, sourced, and managed just like any other component of SOA it is not a one-time 'fire-and-forget' decision. Hence, policies need to be enacted to ensure that

- A technology foundation (often termed Strategic SOA Platform or End-to-End SOA Platform) that provides messaging, security, and other services is centrally funded and leveraged by all projects.
- A governance platform that is part of the SOA platform to enable the automation of policies where possible.
- Consensus is built regarding the migration of legacy systems and platforms to SOA technologies.
- SOA platform enhancements coincide with the project portfolio plan and business service portfolio plan.
- The design and implementation of shared foundation/utility services are a part of SOA infrastructure.

Many architecturally mature organizations today have organizations and processes devoted to software and hardware governance, and these need to be leveraged for building out a SOA infrastructure.

3.3.3 Information

Developers often create service interfaces that perpetuate poor data access methods, which negate the benefits of creating services that share data. To give service-oriented applications a strong foundation, data quality and interoperability issues must be addressed. The goals of any SOA initiative, then, should be to create data standards to overcome disparities in data representations across legacy, ERP and other systems, and to create a set of services that becomes the authoritative way to access high-quality enterprise data. An enterprise data environment should be able to do the following:

- Create single logical sources for key enterprise entities such as customers or products.
- Eliminate custom interfaces and proprietary data formats.
- Improve data quality across the enterprise.

- Enforce data standards in the data services layer.
- Make data readily discoverable, accessible, and interoperable.
- Realize policy-driven security for data services.

The specific data governance issues that need to be addressed by an enterprise data management function include:

- Defining data ownership and stewardship, including roles and responsibilities for data consumers and producers.
- Setting up a data services architecture.
- Establishing policies and guidelines for adhering to the data standards chosen by the enterprise.
- Mandating the use of specific schemas as the format for exchange master data.
- SOA Governance – Framework and Best Practices.
- Establishing processes for exceptions, changes to standards, version management, and so on.
- Mandating the use of specific data services as the single source of truth for the data families they serve.
- Mandating policies to ensure that data services conform to data quality metrics.
- Defining and enforcing the security policies that are applied to data services.
- Defining service-level agreements (SLAs) to which enterprise data services must conform.

3.3.4 People

Adopting SOA requires more than just a technology shift. Policies to encourage desirable behavior among employees must be part of SOA governance. Specific areas that need to be considered include:

- Assigning and empowering employees who are responsible for driving process improvement, often called *process officers.*
- Developing the skills necessary for architecting, building, testing, and deploying services and service-oriented applications.
- Creating incentives to encourage the building of sharable services and the reuse of existing services.
- Forming an enterprise architecture group to drive adoption of EA disciplines and SOA in particular.

- Creating a group that is specifically tasked with governing the SOA road map.

Typically, the SOA governance group consists of representatives from EA, the different lines of business, and finance. Failure to address organizational and change management issues will lead to slow SOA adoption that lacks coherence, because employees are not empowered and are not held accountable for delivering on SOA benefits.

3.4 Governance Policies

Policies set the goals that we use to direct and measure success. Without policies, there is no Governance. Policy Makers like IT Managers, Architects, Project Leaders, and Application Development Leaders are struggling to define, configure, and assign policies in a way that allows IT development teams to easily and transparently adhere to policies and practices. As a result, each team creates services in a slightly differently way. This sacrifices interoperability, manageability, security, and the other benefits of SOA. Policies need to address the overall impact to the business of the Services that are being created and deployed. Policies need to create a strong connection between the business and technology. Companies need the ability to associate their business policies, technical policies and actual implementation in a transparent fashion.

Policies are technical and business requirements that aim to create a common utilized language of information and process. Deployed in SOA, policies need to address the very distributed, asynchronous, and heterogeneous nature of the SOA environment. At the highest level, SOA is supposed to deliver the two big A-words: Agility and Alignment. Otherwise, SOA achieves those goals by replacing rigid, monolithic applications and multiple instances of redundant, static software assets with a consolidated matrix of dynamic, reusable services. These services can be shared across the enterprise, and assembled and re-assembled into a variety of applications. An effective SOA governance is needed to make that happen, and policies are the nuts-and-bolts of governance.

3.4.1 A Question of Access

Access is a straightforward issue, a service or any other software asset is of little value if no one can get to it to use it. The appropriate SOA governance policies must be created and enforced, and the right tools put in place, to both guarantee and control service access.

3.4.2 The Right Stuff

In SOA, usability is synonymous with reusability, which itself is one of the defining characteristics of a service. The motivation for service-enabling some chunk of functionality is to allow it to be consumed by any application that requires it. Reusability, therefore, is something else that SOA governance must enforce. That means creating and applying the policies necessary to insure that the right services are developed the right way at the right time using, whenever possible, other software assets available in the inventory.

3.4.3 Performance Anxiety

Governance over reusability must also extend to service performance in the operational environment. SOA governance must include the means to monitor, measure, and report on service performance in order to insure service quality.

3.4.4 Use It Or Lose It

While service reusability is vitally important, it cannot insure reuse. Even the most accessible, reusable services may end up collecting dust simply because no SOA governance policies are in place to make sure that reusable stuff gets reused. If a SOA without services is worthless, a SOA with services that are not being used is a highly effective cash incinerator. So the policies that represent the application of governance to SOA must accomplish at least two objectives:

1. Ensure that the organization has services to use.
2. Ensure that the organization uses the services it has.

3.4.5 Policies

Policies might start at the business level:

- Projects must comply with Internal Architecture guidelines.
- Security and regulatory compliance policy reviews are mandatory for all IT projects.

Policies could represent more specific regulatory compliance issues:

- Patient personal identifiable information must be communicated and stored securely (HIPPA).
- All financial transactions must provide traceability and tamper proof mechanisms for mandatory audit records (SOX).

Project outsourcing initiative might represent its requirements as:

- Project must have interoperability policy passing score better than 80% during the acceptance testing.
- Any project failing key security policies must be scheduled for a manual review.
- Outsourcer must provide signed acceptance checklist to the project manager prior to the acceptance testing.

These higher level policies will often need to be translated to highly technical policies that can be effectively enforced by active policy enforcement tools.
Information security examples:

- Messages must contain an authorization token.
- Password element lengths must be at least 6 characters long and contain both numbers and letters.
- Every operation message must be uniquely identified and digitally signed.

There are also design related technical policies that are needed to ensure interoperability and reuse:

- Do not use RPC encoded style Web services.
- Do not use Solicit-Response style of Web service operations.
- Do not use XML 'any Attribute' wildcards.

If any one of these sets of policies is not followed, the impact on company operations and bottom line can be enormous.

3.5 Governing Analysis Process

3.5.1 Limitation of BPM, EA, and OOAD

Experience from early SOA implementation projects suggests that existing development processes and notations such as OOAD, EA, and BPM only cover part of the requirements needed to support the SOA paradigm. While the SOA approach reinforces well-established, general software architecture principles such as information hiding, modularization, and separation of concerns, it also adds additional themes such as service choreography, service

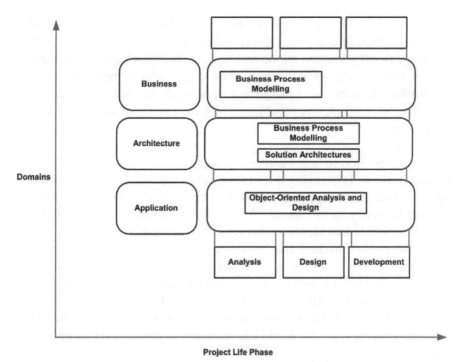

Figure 3.1 BPM, EA, and OOAD positioning.

repositories, and the service bus middleware pattern, which require explicit attention during modeling.

Figure 3.1 illustrates where the existing EA, BPM, and OOAD modeling approaches have their main application areas. It also gives us a good starting point for the following discussion of SOAD. The horizontal axis in the figure is the project lifecycle phase; the vertical one differentiates between the different levels of abstractions or domains, on which modeling activities typically reside.

The SOA vision is accepted rather easily, as its technical foundation is well known. For example, applying general software architecture principles and OO techniques is a valid start in any SOA effort. However, as already stated, the question most frequently asked by early adopters is how to identify the right services. As stated earlier, OOAD, EA, and BPM cannot supply a satisfying answer when applied in isolation from each other.

The OOAD methodology introduced in the seminal books from Booch and Jacobson, provides an excellent starting point in defining SOAs. OOAD is concerned with micro-level abstractions such as classes and individual object instances, although applying OOAD techniques and the Unified Modeling Language (UML) notation on the architectural level has been common practice for many years. As a standalone use case model is frequently created per problem domain, and consequently, application development project, the enterprise-wide big picture gets blurred in many cases. Furthermore, for various reasons the use case models are not always synchronized with their BPM counterparts.

EA approaches, such as Feature-Oriented Domain Analysis (FODA), and Zachman add a city-planning level viewpoint on top of solution architectures, but do not address how enterprise-wide abstraction of quality facilitating re-use and longevity can be found. While BPM approaches such as BPMI do provide an end-to-end view on functional units of work, they typically do not reach into the architecture and implementation domain. For example, until the arrival of languages such as the Business Process Execution Language for Web Services (BPEL), BPM notations were missing operational semantics. Finally, none of the existing disciplines address how existing applications can be enabled for SOA, a top-down process is employed most of the time. Existing systems typically hold large amounts of critical data and business logic, and cannot simply be replaced. Hence, a bottom-up analysis of these systems also has to be conducted in order to investigate wrapping and refactoring strategies. Taking existing applications into account, therefore, leads to a meet-in-the-middle process.

A hybrid SOAD modeling approach is required for these reasons. The approach comprises elements from OOAD, BPM, and EA in a best-of-breed fashion, and complements them with certain innovative elements. Figure 3.2 illustrates the SOAD assets for this new approach:

Evolving enterprise applications and IT infrastructure into a SOA can be a major undertaking, affecting multiple lines of business and organizational units. Therefore, EA frameworks and reference architectures such as The Open Group Architecture Framework (TOGAF) should be applied, striving for architectural consistency between individual solutions. According to past experience, most existing EA frameworks have limitations in one or more areas. Furthermore, many enterprise-level reference architectures and frameworks are rather generic, and do not reach down to the design domain. Such high-level architectures fail to give concrete, tactical advice for solution architects and developers that lead to a fundamental gap between enterprise and solution architectures frequently is the consequence.

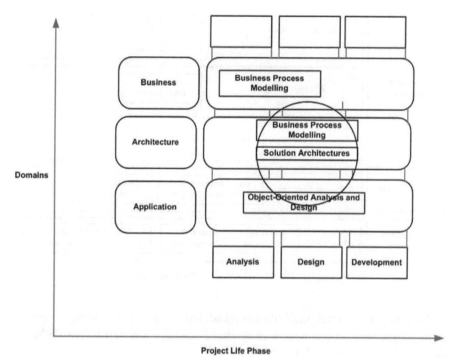

Figure 3.2 SOAD and its ingredients: OOAD, BPM, and EA.

All existing BPM approaches can be leveraged as a starting point for SOAD, however, they have to be amended with additional techniques for deriving candidate services and their operations from the process models. Furthermore, process modeling in SOAD must be synchronized with design-level use case modeling [3].

3.5.2 The SOAD Service Definition Hierarchy

In the process area using the Rational Unified Process (RUP) recognized as one of the leading OOAD processes to support the analysis and design of iterative software development utilizing a UML model is of primary value. However, RUP has the principles of OOAD as its foundation, and therefore, does not lend itself easily to be aligned to SOA design. From the viewpoint of RUP, the architecture of a system is the structure of its major components interacting via defined interfaces. Furthermore, these components are composed of decreasingly smaller components down to a class-level of granularity. In contrast, the architecture of the system in a SOA gen-

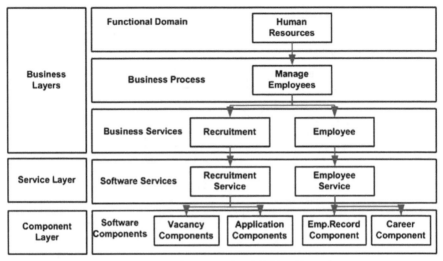

Figure 3.3 The SOAD service definition hierarchy.

erally comprises stateless, fully-encapsulated, and self-describing services satisfying a generic business service that is closely mapped to the BPM, as demonstrated in Figure 3.3.

These services might be composed of a number of collaborating or orchestrated services. This does not exclude the OO viewpoint adopted by RUP, but rather implements another layer of abstraction above it. This super-layer is to encapsulate components that are designed as RUP artifacts within a formal, cross-layer interface structure [2].

3.5.3 Elements of SOAD

The elements that are primary for the SOAD architecture are identified as: Process and notation have to be formally or semi-formally defined, just like in any other project or design methodology. SOAD does not have to start from scratch. By selecting and combining OOAD, BPM, and EA elements, extra elements can then be defined if needed.

There must be a structured way to conceptualize services: OOAD gives us classes and objects on the application level, while BPM has event-driven process models. SOAD has the issue of gluing them together. The method is no longer use case-oriented, but driven by business events and processes. Use case modeling comes in as a second step on a lower level.

The method includes syntax, semantics, and policies. This is required for ad hoc composition, semantic brokering, and runtime discovery.

SOAD must provide well-defined, quality factors and best practices. The roles question raised by BPEL must be answered. SOAD must facilitate end-to-end modeling and have comprehensive tool support. If SOA is supposed to bring flexibility and agility to the business, the same should be expected from its supporting method, spawning from the business to the architecture and application design domains [2].

3.5.4 Quality Factors

Some general principles or quality factors can already be identified and act as a design baseline within SOAD they include:

- Well-crafted services bring flexibility and agility to the business, they facilitate ease of reconfiguration and reuse through loose coupling, encapsulation, and information hiding.
- Well-designed services are meaningful and applicable for more than enterprise application, dependencies between services are minimized and explicitly stated.
- Services abstractions are cohesive, complete, and consistent. A frequently stated assumption is that services are stateless, this statement would be weakened to request services to be as stateless as possible in the given problem domain and context.
- The service naming is understandable for domain experts without deep technical expertise.
- In a SOA, all services follow the same design philosophy and interaction patterns, the underlying architectural style can easily be identified.
- The development of the services and service consumers requires only basic programming language skills in addition to domain knowledge, middleware expertise is only required for a few specialists, that in an ideal world, work for tool and runtime vendors, and not for the companies crafting enterprise applications as SOAs.

3.5.5 Service Identification and Definition

Top-down, business-level modeling techniques such as CBM can provide the starting point for the SOA modeling activities. But a SOA implementation rarely starts on the green field, that creating a SOA solution will almost always involve integrating existing legacy systems by decomposing them into services, operations, business processes, and business rules. Existing applications and vendor packages are factored into sets of discrete services that represent groups of related operations. Business processes and rules are abstracted from the applications into a separate BPM, managed by a business choreography model. All OOAD techniques can be applied in relationship to identifying and defining a service; however, a higher viewpoint needs to be taken. Furthermore, as SOAs aim higher than the classical development project, there is room for additional creative thinking.

3.5.6 Direct and Indirect Business Analysis

BPM and direct requirements analysis through stakeholder interviews and CBM are an obvious and well-suited way of identifying candidate services. When mining for candidate services, product managers and other business leaders should be interviewed.

3.5.7 Service Granularity

As SOA does not equal Web services and SOAP, different protocol bindings can be used to access services residing on different levels of abstraction. Another option is the bundling of several related services into coarser-grained service definitions, which is a variation of the facade pattern.

3.5.8 Naming Conventions

An enterprise-wide naming scheme should be defined. A simple example would be to recommend always assigning a service with a noun, and its operations with verbs. This best practice originates from the OOAD space.

3.5.9 First Genuine SOAD Elements

In addition to the combination of OOAD, BPM, and EA techniques, there are several important SOAD concepts and aspects, which have yet to be fleshed out:

- Service categorization and aggregation.
- Policies and aspects.
- Meet-in-the-middle processes.
- Semantic brokering.
- Service harvesting and knowledge brokering.

3.5.9.1 Service Categorization and Aggregation

Services have different uses and purposes; for example, software services can be distinguished from business services. Furthermore, atomic services can be orchestrated into higher level, full-fledged services. Service composition is simplified by executable models, this is something traditional modeling tools and methods do not deal with.

3.5.9.2 Policies and Aspects

A service has syntax, semantics, and QoS characteristics that all have to be modeled, formal interface contracts have to cover more than the Web Services Description Language (WSDL) does. Therefore, the WS-Policy framework is an important related specification.

Business traceability is a desirable quality, in addition to the well-established principle of architectural traceability: it should be possible to directly link all runtime artifacts directly to the language a non-technical domain expert can understand. This is particularly important for abstractions directly exposed at the business and service layer. The Sarbanes-Oxley (SOX) act is an example for a business driver requiring this kind of business traceability.

3.5.9.3 Process: Meet-in-the-Middle

There are no green field projects in the real-world as legacy applications that has to be always have to be taken into account. Therefore, a meet-in-the-middle approach is required, rather than pure, top-down or bottom-up process.

The bottom-up approach tends to lead to poor business-service abstractions in case the design is dictated by the existing IT environment, rather than existing and future business needs. On the other hand, top-down processing might cause insufficient, non-functional requirement characteristics, and compromise other architecture quality factors as well as provide impedance mismatches on the service and component layer.

3.5.9.4 Semantic Brokering

In any SOA context, a formal interface contract for the invocation syntax is important. The semantics issue (the meaning of parameters and so forth) has to be solved as well (domain modeling). This is key in any business-to-business (B2B) and dynamic invocation scenario. Such scenarios are cornerstones of the SOA vision of being flexible and agile in response to the new business needs in a world of mergers and acquisitions, business transformation, globalization, and so forth.

3.5.9.5 Service Harvesting and Knowledge Brokering

This is a knowledge management and lifecycle issue. Services should be identified and defined with reuse as one of the main driving criteria of the SOA in mind. If a component (or service) has no potential for reuse, then it should probably not be deployed as a service. It can be connected to another service associated with the enterprise architecture, but will not be a service in its own right. However, even if reuse is planned for right from the beginning, the process still has to formalize the service harvesting process. Service usage by more than one consumer is an explicit SOA design goal. A build-time service registry, for example, an enterprise-wide UDDI directory can be part of the answer.

SOAD will require enhancements to existing software engineering methods, further improving their usability and applicability to enterprise application development projects. Related best practices and patterns will evolve over time. It is recognized that UML will continue to dominate as the notation of choice on the process side enhancements will likely be required to satisfy the wider scope of SOAD.

The next step on the road to a complete SOAD method will be to define the required end-to-end process and notation, review the required roles on engagements and their responsibilities, and continue to validate the proposed approach on projects.

3.6 Governing Run-Time Behavior

Interoperability and service reuse are among the major promises of service-oriented architecture (SOA). Yet interoperability and reuse can only be fully realized when everyone is working on the same page. Hence, SOA has been a key driver in the increasing emphasis on, and interest in, governance in recent years. Leading the charge for governance have been enterprise architects, who know quite well that for SOA systems to deliver value, there must be control in areas ranging from service design and deployment processes, to granular items such as schemas and WSDL creation.

Formerly, given the early stage of SOA technology and practices, it made sense that organizations implementing such systems focused primarily on these areas, especially since most companies were still in the development and design phase. Today, however, with SOA services now in production

within many organizations, system architects are realizing that the most critical area for control and governance is now runtime. Data point after data
point has demonstrated that many SOA implementations are just not working
in production as designed or expected. Problems range from service interrup-
tions to entire business processes failing and security and compliance risks
that generate costly delays and lengthy triage cycles.

3.6.1 The Four Stages

Runtime governance can be divided into four primary areas: process, measurement, enforcement, and feedback. Process comes first because if it is
compromised, circumvented, or not adhered to, there can be no effective
control.

3.6.1.1 Process

While a great deal of process is employed in the SOA pre-production side,
active governance kicks in when an application is migrated from development
and into production. It is at this point that runtime governance can detect
and report if services or consumers in production are adhering to governance
guidelines.

3.6.1.2 Measurement

While significant governance work and planning occur in design and development, what is critical to governance is what occurs in the runtime environment
and, more to the point, knowing what is going on across your SOA during
runtime.

3.6.1.3 Enforcement

End-to-end visibility and control over business processes are critical to enforcing business and IT rules, reporting on them, and having the ability
to do something about them in real time. Specifically, when governance
guidelines are enforced, a runtime system can dynamically react to business
opportunities or IT issues to directly impact the bottom line.

3.6.1.4 Feedback

Systematically tracking governance infractions and tracing their causes facilitate a lifecycle approach in which organizations are able to quickly fix and
address breaches upstream. Runtime governance plays a crucial role as the
last line of defense and is designed to protect the company and the IT system.

3.7 Summary

Governance is a critical piece of the overall running strategy of any organization, generating the process adherence, measurement, enforcement, and feedback that are necessary for an effective lifecycle approach. In planning a proper SOA deployment, it is important to have runtime governance 'baked-into' the development cycle early in order to avoid having the last line of defense become the only line of defense. With rogue service elimination, runtime governance is not just a 'nice-to-have' point solution for replacing a manual process. It is a mandatory requirement for proper SOA security and compliance because it helps to eliminate risk associated with non-compliance and invisible interdependencies that can lead to catastrophic SOA failures, missed business opportunities, and potentially devastating violations of government regulations.

References

1. Jan Pavlovič. Process-Oriented Modeling and Infrastructure for Remedial Decision Support System. Masaryk University, 2009.
2. F. Leymann, D. Roller, and M.T. Schmidt. Web Services and Business Process Management. *IBM Systems Journal*, 41(2), 2002.
3. Olaf Zimmermann, Pal Krogdahl, and Clive Gee. Elements of Service-Oriented Analysis and Design, SOA and Web Services. IBM, http://www.ibm.com/developerworks/webservices/library/ws-soad1, 2004.
4. Qiu-Xia Zhang and Xiao-Guang Zhang. Modeling Service-Oriented Enterprise Architecture. *SOA World Magazine*, http://Soa.Sys-Con.Com/Node/1030392, 2009.

4

SOA and Business Process Management

4.1 Business Process Management Concepts

Business Process Management (BPM) is a management approach focusing on aligning all aspects of an organization with the wants and needs of clients. It is a holistic management approach [1] that promotes business effectiveness and efficiency while striving for innovation, flexibility, and integration with technology. Business process management attempts to improve processes continuously. It could therefore be described as a 'process optimization process'. It is argued that BPM enables organizations to be more efficient, more effective and more capable of change than a functionally focused, traditional hierarchical management approach.

4.1.1 Overview

A business process is a series or network of value-added activities, performed by their relevant roles or collaborators, to purposefully achieve the common business goal. These processes are critical to any organization as they generate revenue and often represent a significant proportion of costs. As a managerial approach, BPM considers processes to be strategic assets of an organization that must be understood, managed, and improved to deliver value added products and services to clients. This foundation is very similar to other Total Quality Management or Continuous Improvement Process methodologies or approaches. BPM goes a step further by stating that this approach can be supported, or enabled, through technology to ensure the viability of the managerial approach in times of stress and change. In fact, BPM is an approach to integrate a 'change capability' to an organization – both human and technological. BPM can be discussed from two points of view: people and/or technology.

The idea of (business) process is as traditional as concepts of tasks, department, production, outputs. The current management and improvement approach, with formal definitions and technical modeling, has been around since the early 1990s .Although the initial focus of BPM was on the auto-mation of business processes with the use of information technology, it has since been extended to integrate human-driven processes in which human interaction takes place in series or parallel with the use of technology. For example (in workflow systems), when individual steps in the business process require human intuition or judgment to be performed, these steps are assigned to appropriate members within the organization.

More advanced forms such as human interaction management are in the complex interaction between human workers in performing a workgroup task. In this case, many people and systems interact in structured, ad-hoc, and sometimes completely dynamic ways to complete one to many transactions.

BPM can be used to understand organizations through expanded views that would not otherwise be available to organize and present. These views include the relationships of processes to each other which, when included in the process model, provide for advanced reporting and analysis that would not otherwise be available. BPM is regarded by some as the backbone of enterprise content management.

Because BPM allows organizations to abstract business process from technology infrastructure, it goes far beyond automating business processes (software) or solving business problems (suite). BPM enables business to respond to changing consumer, market, and regulatory demands faster than competitors – creating competitive advantage.

Most recently, technology has allowed the coupling of BPM to other methodologies, such as Six Sigma. BPM tools now allow the user to:

- *Define* – Baseline the process or the process improvement.
- *Model* – Simulate the change to the process.
- *Analyze* – Compare the various simulations to determine an optimal improvement.
- *Improve* – Select and implement the improvement.
- *Control* – Deploy this implementation and by use of user defined dash-boards monitor the improvement in real time and feed the performance information back into the simulation model in preparation for the next improvement iteration.

This brings with it the benefit of being able to simulate changes to your business process based on real life data (not assumed knowledge). Also, the coupling of BPM to industry methodologies allows users to continually

Figure 4.1 Business Process Management (BPM).

streamline and optimize the process to ensure that it is tuned to its market need.

4.1.2 BPM Lifecycle

Business process management activities can be grouped into five categories: design, modeling, execution, monitoring, and optimization.

4.1.3 Design

Process design encompasses both the identification of existing processes and the design of 'to-be' processes. Areas of focus include representation of the process flow, the actors within it, alerts & notifications, escalations, Standard Operating Procedures, Service Level Agreements, and task hand-over mechanisms.

A good design reduces the number of problems over the lifetime of the process. Whether or not existing processes are considered, the aim of this step is to ensure that a correct and efficient theoretical design is prepared.

The proposed improvement could be in human-to-human, human-to-system, and system-to-system workflows, and might target regulatory, market, or competitive challenges faced by the businesses.

4.1.4 Modeling

Modeling takes the theoretical design and introduces combinations of variables (e.g., changes in rent or materials costs, which determine how the process might operate under different circumstances).

It also involves running 'what-if analysis' on the processes: 'What if I have 75% of resources to do the same task?' 'What if I want to do the same job for 80% of the current cost?'

4.1.5 Execution

One of the ways to automate processes is to develop or purchase an application that executes the required steps of the process; however, in practice, these applications rarely execute all the steps of the process accurately or completely. Another approach is to use a combination of software and human intervention; however this approach is more complex, making the documentation process difficult.

As a response to these problems, software has been developed that enables the full business process (as developed in the process design activity) to be defined in a computer language which can be directly executed by the computer. The system will either use services in connected applications to perform business operations (e.g. calculating a repayment plan for a loan) or, when a step is too complex to automate, will ask for human input. Compared to either of the previous approaches, directly executing a process definition can be more straightforward and therefore easier to improve. However, automating a process definition requires flexible and comprehensive infrastructure, which typically rules out implementing these systems in a legacy IT environment.

Business rules have been used by systems to provide definitions for governing behaviour, and a business rule engine can be used to drive process execution and resolution.

4.1.6 Monitoring

Monitoring encompasses the tracking of individual processes, so that information on their state can be easily seen, and statistics on the performance of one or more processes can be provided. An example of the tracking is being able to determine the state of a customer order (e.g. ordered arrived, awaiting delivery, invoice paid) so that problems in its operation can be identified and corrected.

In addition, this information can be used to work with customers and suppliers to improve their connected processes. Examples of the statistics are the generation of measures on how quickly a customer order is processed or how many orders were processed in the last month. These measures tend to fit into three categories: cycle time, defect rate and productivity.

The degree of monitoring depends on what information the business wants to evaluate and analyze and how business wants it to be monitored, in real-time, near real-time or ad-hoc. Here, Business Activity Monitoring (BAM) extends and expands the monitoring tools in generally provided by BPMs.

Process mining is a collection of methods and tools related to process monitoring. The aim of process mining is to analyze event logs extracted through process monitoring and to compare them with an a priori process model. Process mining allows process analysts to detect discrepancies between the actual process execution and the a priori model as well as to analyze bottlenecks.

4.1.7 Optimization

Process optimization includes retrieving process performance information from modeling or monitoring phase; identifying the potential or actual bottlenecks and the potential opportunities for cost savings or other improvements; and then, applying those enhancements in the design of the process. Overall, this creates greater business value.

4.1.8 BPM Technology

There are four critical components of a BPM suite:

1. Process Engine – a robust platform for modeling and executing process-based applications, including business rules.
2. Business Analytics – enable managers to identify business issues, trends, and opportunities with reports and dashboards, and react accordingly.

3. Content Management – provides a system for storing and securing electronic documents, images, and other files.
4. Collaboration Tools – remove intra- and interdepartmental communication barriers through discussion forums, dynamic workspaces, and message boards.

BPM also addresses many of the critical IT issues underpinning these business drivers, including:

- Managing end-to-end, customer-facing processes.
- Consolidating data and increasing visibility into and access to associated data and information.
- Increasing the flexibility and functionality of current infrastructure and data.
- Integrating with existing systems and leveraging emerging Service-Oriented Architectures (SOAs).
- Establishing a common language for business-IT alignment [2].

4.2 The Role of Business Process Management in SOA

In today's competitive environment where companies are merging, consolidating and striving to uncover new growth opportunities, savvy business leaders are recognizing the value that comes from working more closely with IT professionals.

Driven in large part by the growing adoption rates of a service-oriented architecture (SOA) strategy, more and more organizations are realizing that the alignment of IT and business delivers tangible results and significant returns in terms of productivity, competitive advantage and cost savings.

However, to seize these new opportunities and realize the benefits that can be derived from a SOA, companies need to streamline their business processes and eliminate the re-creation of the wheel that too often happens when an organization and its technology resources are locked into silos.

For example, these business processes can include such functions as ordering supplies, reimbursing expenses or booking business travel. When each department or team within a larger organization has its own system to handle these types of business processes, it may prove effective for a smaller subset of the company but is largely ineffective for the entire business. Consider the advantages that can be reaped in terms of productivity and cost savings when there is an agreed-upon approach to filing an expense report or automating travel requests.

Addressing these issues and uncovering ways to automate and improve business processes without requiring additional resources is top of mind for today's organizations as they aim to more effectively and efficiently compete in an ever-changing marketplace. This growing need has led to the rise of the BPM market.

4.2.1 The Growing Demand for Business Process Management

BPM is a discipline combining software capabilities and business expertise to accelerate process improvement and facilitate business innovation. One could argue that BPM is based on the principles of SOA, with both aiming to empower the organization to more quickly respond to changing market conditions that result from planned events such as mergers and acquisitions or external influences such as competitor moves.

Several factors are driving the increased focus on BPM. These include the need to:

- Ensure consistency throughout the company, especially with regard to compliance.
- Optimize processes for maximum efficiency.
- Automate manual processes to reduce time-consuming administrative tasks.
- Integrate complex, redundant processes.
- Mitigate risks through a single, unified view of the organization.

A successful BPM solution will take existing processes, streamline them to meet business goals and, ultimately, impact the bottom and top lines in a positive way. The value of BPM is further evidenced by the results that can be realized from business and IT working more closely together. One of the most significant benefits is the fact that BPM helps to put business process control in the hands of business managers. By providing decision-makers with up-to-date business information, BPM allows them to make better decisions immediately without relying on IT support.

BPM is growing in popularity and is complementary to SOA due to its ability to help make business processes more efficient and effective while enabling an organization to more easily adapt to changing business requirements. BPM based on SOA is technology's response to the growing demand for a flexible business environment that is not hindered by application silos.

When business processes are automated, streamlined and supported by a strong SOA governance framework, BPM can deliver on its promise of transforming IT processes to dynamically adapt to business needs. For these reasons, BPM is being widely embraced. In fact, analysts at IDC state that the BPM tools market will reach $3 billion and above by 2012.

The powerful combination of BPM to streamline business processes within a SOA strategy will help position companies to become industry leaders while ensuring they are poised for continued success. For this to happen, however, business processes must become independent of specific information resources and specific task automation applications. The integration technology must loosely couple the applications and resources that make up the process, otherwise the logic of a process will get hard coded into a particular technology platform, which may be expensive to change and therefore defeat the entire purpose of BPM.

The need to model business processes before they are deployed in a SOA is becoming increasingly important, especially as the demand for BPM continues to rise. First, however, let us understand that BPM is both a management discipline and a technology platform and that modeling is a complementary and critical aspect within a larger BPM strategy.

As a management discipline, BPM replaces traditional views of business based on discrete functional organizations, systems and metrics with those based on cross-functional core processes aligned with high-level business objectives. As a technology platform, BPM provides the set of software tools needed to optimize performance, make abstract performance goals concrete, connect them to process data, automate and monitor process activities and provide a platform for agile performance improvement.

One of the most enticing benefits of BPM is the fact that it delivers greater flexibility throughout the organization while using and reusing existing technology investments. Once you realize this, you gain a greater understanding of how modeling complements the management and technology views of BPM while helping to drive greater business results through a SOA.

Organizations realize that the role of BPM in a SOA cannot be overlooked. Just as modeling business processes before deployment is a vital first step in a SOA, BPM is equally important because of its capabilities to evolve business processes from merely hard-wired automated functions to delivering the much-needed business flexibility within those business processes. The added benefit of BPM is that by eliminating the hard wiring, business processes can be continuously improved and easily shared throughout the organization.

As companies look to the power of SOA to drive business results, they are realizing the value of modeling is one of the most critical steps to SOA success. This is primarily due to the fact that modeling helps ensure that internal processes are directly aligned with strategy and goals before they are implemented.

More specifically, modeling helps organizations fully visualize, comprehend and document business processes in order to close the gap that exists between an organization's lines of business and its understanding of the business drivers. Given that a business process is a defined set of activities leading to specific results, modeling provides the added assurance that best practices are well documented and communicated throughout the organization before deployment.

For example, business analysts can use modeling to define alternative scenarios, differing in resource allocation, branching assumptions at decision points in the flow, and other parameters, and see which alternative results in the lowest cost, fastest average cycle time, lowest percentage of service level agreements violations or other optimum business measures. In addition, this simulation can help reveal bottlenecks in the process, allowing new alternative scenarios to be analyzed, and resulting in significant time and cost savings before they are implemented throughout the SOA.

Together, BPM and SOA help facilitate the next phase of the business process evolution. The evolution is occurring now because of the heightened need for enterprises to compete more effectively by adapting to market changes faster, continuously improving efficiencies and streamlining collaboration across traditionally siloed departments. This business process evolution is resulting in a wider adoption of best practices throughout the company while eliminating costly redundancies. For example, once an organization shares business processes beyond a team or even a department, the processes cannot help but improve because there are a greater number of parties focused on overall productivity and organizational excellence.

For BPM to be successful and valuable to the enterprise, the speed and agility of IT organizations implementing and integrating the process automation components must match the speed and agility of business analysts redesigning the process.

Just as BPM capabilities needed to evolve over time to add flexibility to process design, so too do application integration systems need to evolve to automate the new flexibility processes in the real world. This integration evolution requires the ability to create independence between process and service implementation, and to remove the tight coupling between a specific integration technology and individual business applications. This is where SOA comes in because it provides the technical ability to create that process implementation independence.

Making those value changes to processes requires integration between existing and future applications that automate specific business functions. Automation only becomes flexible if it can be reused and reintegrated in a dynamic manner. A standards-based SOA infrastructure is designed to deliver the automation flexibility, and Web services is designed to provide the technology standards to make dynamic integration a reality across departmental and enterprise boundaries.

SOA assumes that IT portfolio items will change over time. SOA infrastructure assumes that business processes dictating how and when those items will be used and communicate with each other will change over time. This process independence from how specific automation components are implemented helps make technology resources as flexible as the process models provided by the BPM solution. Enterprises may then fully merge process improvement efforts with technology resource management. When both are done together, enterprises may achieve dramatic improvements in market capture, cost-effectiveness and profitability.

4.3 Working with Dynamic BPM and SOA Environments

Dynamic BPM is the ability to support process change at any time, with very low latency. Dynamic BPM implies that your business processes are more agile and flexible so it can easily adapt dynamically based on business needs.

4.3.1 Dynamic BPM Vs Traditional BPM

To give a short overview, the initial SOA BPM solutions focused on streamlining business processes, aligning Business and IT content, process automation and standardizing IT interfaces. These processes were not designed with agility and dynamicity in mind. Now, based on dynamicity associated with today's business environment, business needs to sense environmental changes and adapt quickly and respond appropriately to these changes. That is where Dynamic BPM comes into play.

We need to have the right platform that provides an end to end capability to model, assemble, deploy and manage these dynamic business processes.

IBM WebSphere Dynamic Process Edition offered by IBM is a key IBM BPM Suite element, which provides this built-in support for adapting, responding dynamically to change and provides the BPM enabled by SOA foundational capabilities for modeling, simulating, deploying and monitoring end-to-end dynamic business processes [2].

4.3.2 The Benefits of Dynamic Business Processes

Ensuring an appropriate environment exists to enable more dynamic processes, improved control and visibility, and timely accurate information requires the implementation of a Business Process Management System (technology enabler).

The instantaneous benefit that a BPMS will deliver is the automation of business processes. This in turn results in a structured and governable way of working. This means that Information workers will only need to interact with a specific process when they specifically need to. For the activities where they are not involved the BPMS will take care of the process activities, notifications, escalations and executing business rules in the background.

4.3.3 Visibility & Control

Dynamic BPM provides visibility into the real-time status of an entire end to end process and any related activities for information workers and managers. This means that a more proactive approach can become common practice. This level of visibility proves highly beneficial when deadlines are approaching or work items are overdue. Furthermore, executives achieve insight into performance standards such as how long it takes for an employee to complete a task, how many activities an employee can successfully undertake at once and how often an employee delegates tasks, thereby enabling better business decisions in the future.

4.3.4 Accountability

Dynamic BPM supports a culture of organizational and personal accountability by tracking and auditing individual's turnaround time and quality of work. It is also possible to do sampling of specific process tasks to ensure that compliance for critical tasks and processes are in place.

4.3.5 Improved Productivity

In addition to providing executives and managers with increased visibility and control, Dynamic BPM also drives knowledge worker productivity by capturing and interpreting the business context of each task and proactively providing the worker with the content required to complete it.

4.3.6 Relating to Business Process Management

The occurrence of inappropriate and untimely emails, i.e. emails sent out of context but in relation to a specific process task, is often a symptom of an organization that has broken and inefficient processes or lack of enforcement of existing business processes.

Many organizations believe that if they have well modeled processes (traditional BPM), they will be well understood and therefore followed rigorously by the average employee. This is Flawed logic. Organizations are continually undergoing change and therefore the aforementioned Traditional BPM concept is not practical. It does not scale and is hardly ever implemental without vast amounts of bureaucracy.

In addition, the complexity of the typical cross functional processes (traditionally manual and paper based) force information workers to 'reinvent' or 'informally optimized' an existing process due to time or resource constraints.

Management and employees have difficulty in making correct decisions due to not having accurate and timely information.

Information is of devoid of business context (not seen in as part of a business process) and therefore may lead inconsistent decisions. Planning is often difficult due to 'broken' processes, poor working practices and inconsistent information. IT systems that support the various business processes act as inhibitors rather than enablers as they are typically designed deployed and managed as information islands or stove pipes. Enforcing business rules, policies and procedures is an arduous task due to poor process visibility, accurate real-time information and no integrated or consistent auditing. This dichotomy between 'traditional' BPM and the failure to move to a dynamic BPM model is largely due to organizations not adopting a pragmatic approach to solving high value, high priority business problems [1].

4.3.7 SOA Roles Played by BPM

BPM as a Service Consumer
Services whose operations provide access to enterprise applications:

- Create, update, cancel order.
- Get order status.

BPM as a Service Provider
Services whose operations define and manage business processes:

- Start process.

- Start, complete activity.
- Get activity status.

4.4 Coordinating BPM, SOA and Web Services

The internet has created universal connectivity. The extended enterprise uses the internet to connect to its customers, partners and suppliers. Web services have forever changed the way we think about enterprise application and architectures. Web services allow language, platform and location independent connectivity using open universal standards like HTTP. Web services allow heterogeneous applications to talk to each other without expensive EAI software. Web services enable existing application to be exposed as services. SOA is a collection of loosely coupled Web services that communicate with one another. SOA where business logic and business processes are exposed to other processes and software through standard services. BPM and Web services compliment each other. Web services can be used by the BPM system as a way to integrate other IT applications in a process centric way. Secondly, the BPM system itself can be exposed as a Web service and other applications and systems can initiate a process through the BPM system. BPM and Web services compliment each other. Web services can be used by the BPM system as a way to integrate other IT applications in a process centric way [3].

4.4.1 A Business Process Calls Internal and External Web Services

A business process calls internal and external Web service shows a business process invoking internal and external Web services. External Web services are invoked over the internet and can be used to collaborate with external partners and suppliers. BPM and Web services are a powerful combination and when used together the combination creates mind blowing flexibility, loosely coupled integration and excellent opportunities for collaboration and business process outsourcing.

4.4.2 BPM and Web Services – A Powerful Combination

BPM and Web services will dominate the next few years. This combination creates a rich matrix of process possibilities. BPM and Web services facilitate a new breed of process automation and innovation in the extended enterprise.

The dynamic combination of BPM and Web services can truly create the following process possibilities

- Process Plug and Play.
- Process embedded within processes.
- Universal business processes.
- Specialized business processes.
- Process services.
- Process collaboration.

Web services are becoming a major catalyst for the evolution of BPMS. The ability to dynamically invoke Web services in a synchronous or asynchronous manner from within the business process during process orchestration is enabling the creation of new industries and process innovation. In fact the cost and complexity of BPM solutions can be reduced through the use of Web services. There are certain challenges that need to be kept in mind for complex business processes like

- Coordination of asynchronous communication between the business process and the Web services.
- Long running transactions during process orchestration.
- Correlate message exchanges between the business process and Web services.

EBay is the worlds biggest online auction site. It has converted itself to the biggest flea market of the planet. Nearly 40% of the product listing are generated by Web services. By making eBay accessible through Web services, business are empowered to leverage eBay's trading platform and auction process. Businesses are embedding eBay's auction Web service into their own business processes like traditional sales, delivery and procurement processes. This new plug and play process and Web service architecture is generating efficiencies and profits for forward looking companies.

Processes managed enterprises are creating new industries by making available internal business functions as pay-per-use processes and creating new industries.

4.4.3 eBay's Auction Process Exposed as a Web Service

eBay's auction process is exposed as a Web service. Vendor's delivery and procurement processes call eBay's auction process over the internet using Web services. This real time example shows process 'plug and play' and 'pay-per-use' process architecture. Vendors now can embed eBay's super efficient auction process into their own business processes and give eBay a service fee. EBay's auction process communicates back to the vendor's process and the vendor has complete visibility of product auction and complete control. BPM and SOA together will create virtual enterprises like eBay. The point is organizations do not have to create everything on their own. They can create innovative processes that contain sub processes created and managed by others.

Using BPM and Web services in tandem may allow the extended enterprise to extract far more value of internal applications and leverage external business processes.

As BPM moves towards mainstream adoption, we are going to see an evolution of business processes and services. Web services will quietly merge behind the scenes with BPMS, and we are likely to see processes every where. Universal services like currency rate services will be offered and maintained by some company. All other business processes will just consume the currency rate service.

Business process and Web services combination is already changing the business landscape. Business processes will invoke Web services, but it might be the end consumer interested in the results of the Web service. Dell might use Web service from UPS in its direct process model to offer its consumer package tracking. Dell consumers go to UPS we site to track their packages once Dell gives them a confirmation number. Such kind of process handoff and process specialization is going to be the norm going forward.

4.4.4 Service-Oriented Architecture (SOA) and Web Services: The Road to Enterprise Application Integration (EAI)

Most enterprises have made extensive investments in system resources over the course of many years. Such enterprises have an enormous amount of data stored in legacy enterprise information systems (EIS), so it is not practical to discard existing systems. It is more cost-effective to evolve and enhance EIS. But how can this be done? SOA provides a cost-effective solution.

- SOA is not a new concept. Sun defined SOA in the late 1990s to describe Jini, which is an environment for dynamic discovery and use of services over a network. Web services have taken the concept of services introduced by Jini technology and implemented it as services delivered over the Web using technologies such as XML, Web Services Description Language (WSDL), Simple Object Access Protocol (SOAP), and Universal Description, Discovery, and Integration (UDDI).

SOA is emerging as the premier integration and architecture framework in today's complex and heterogeneous computing environment. Previous attempts did not enable open interoperable solutions, but relied on proprietary APIs and required a high degree of coordination between groups. SOA can help organizations streamline processes so that they can do business more efficiently, and adapt to changing needs and competition, enabling the software as a service concept. eBay for example, is opening up its Web services API for its online auction. The goal is to drive developers to make money around the eBay platform. Through the new APIs, developers can build custom applications that link to the online auction site and allow applications to submit items for sale. Such applications are typically aimed at sellers, since buyers must still head to ebay.com to bid on items. This type of strategy, however, will increase the customer base for eBay. SOA and Web services are two different things, but Web services are the preferred standards-based way to realize SOA.

4.4.5 Service-Oriented Architecture

SOA is an architectural style for building software applications that use services available in a network such as the Web. It promotes loose coupling between software components so that they can be reused. Applications in SOA are built based on services. A service is an implementation of a well-defined business functionality, and such services can then be consumed by clients in different applications or business processes.

SOA allows for the reuse of existing assets where new services can be created from an existing IT infrastructure of systems. In other words, it enables businesses to leverage existing investments by allowing them to reuse existing applications, and promises interoperability between heterogeneous applications and technologies. SOA provides a level of flexibility that was not possible before in the sense that:

- Services are software components with well-defined interfaces that are implementation-independent. An important aspect of SOA is the separation of the service interface (the what) from its implementation (the how). Such services are consumed by clients that are not concerned with how these services will execute their requests.
- Services are self-contained (perform predetermined tasks) and loosely coupled (for independence).
- Services can be dynamically discovered.

- Composite services can be built from aggregates of other services.

SOA uses the find-bind-execute paradigm as shown in Figure 4.1. In this paradigm, service providers register their service in a public registry. This registry is used by consumers to find services that match certain criteria. If the registry has such a service, it provides the consumer with a contract and an endpoint address for that service.

SOA-based applications are distributed multi-tier applications that have presentation, business logic, and persistence layers. Services are the building blocks of SOA applications. While any functionality can be made into a service, the challenge is to define a service interface that is at the right level of abstraction. Services should provide coarse-grained functionality.

4.4.6 Realizing SOA with Web Services

Web services are software systems designed to support interoperable machine-to-machine interaction over a network. This interoperability is gained through a set of XML-based open standards, such as WSDL, SOAP, and UDDI. These standards provide a common approach for defining, publishing, and using Web services.

Sun's Java Web Services Developer Pack 1.5 (Java WSDP 1.5) and Java 2 Platform, Enterprise Edition (J2EE) 1.4 can be used to develop state-of-the-art Web services to implement SOA. The J2EE 1.4 platform enables you to build and deploy Web services in your IT infrastructure on the application server platform. It provides the tools you need to quickly build, test, and deploy Web services and clients that interoperate with other Web services and clients running on Java-based or non-Java-based platforms. In addition, it enables businesses to expose their existing J2EE applications as Web services. Servlets and Enterprise JavaBeans components (EJBs) can be exposed as Web services that can be accessed by Java-based or non-Java-based Web service clients. J2EE applications can act as Web service clients themselves, and they can communicate with other Web services, regardless of how they are implemented.

4.4.7 Web Service APIs

The Java WSDP 1.5 and J2EE 1.4 platforms provide the Java APIs for XML (JAX) that are shown in Table 4.1.

With the APIs described in Table 4.1, you can focus on high-level programming tasks, rather than low-level issues of XML and Web services. In

Table 4.1 Java APIs for XML (JAX) provided by J2EE 1.4. (Source: Ed Ort, 2005)

API	Description
Java API for XML Processing (JAXP) 1.2	This API lets you process XML documents by invoking a SAX or DOM parser in your application. JAXP 1.2 supports W3C XML Schema.
Java API for XML-based RPC (JAX-RPC) 1.1	This is an API for building and deploying SOAP+WSDL Web services clients and endpoints.
Java APIs for XML Registries (JAXR) 1.0.4	This is a Java API for accessing different kinds of XML registries. It provides you with a single set of APIs to access a variety of XML registries, including UDDI and the ebXML Registry. You do not need to worry about the nitty-gritty details of each registry's information model.
SOAP with Attachments API for Java (SAAJ) 1.2	This API lets you produce and consume messages conforming to the SOAP 1.1 specification and SOAP with Attachments note.
JSR 109: Web services for J2EE 1.0	JSR 109 defines deployment requirements for Web services clients and endpoints by leveraging the JAX-RPC programming model. In addition, it defines standard deployment descriptors using the XML Schema, thereby providing a uniform method of deploying Web services onto application servers through a wide range of tools.

other words, you can start developing and using Java WSDP 1.5 and J2EE 1.4 Web services without knowing much about XML and Web services standards.

Figure 4.2 illustrates how the JAXR and JAX-RPC APIs play a role in publishing, discovering, and using Web services and thus realizing SOA.

4.4.8 Web Services Endpoints in J2EE 1.4

The J2EE 1.4 platform provides a standardized mechanism to expose Servlets and EJBs as Web services. Such services are considered Web service endpoints (or Web service ports), and can be described using WSDL and published in a UDDI registry so that they can be discovered and used by Web service clients.

Once a Web service is discovered, the client makes a request to a Web service. The Web service processes the request and sends the response back to the client. Consider Figure 4.3, which shows how a Java client communicates with a Java Web service in the J2EE 1.4 platform. Note that J2EE applications can use Web services published by other providers, regardless of how they are implemented. All the details between the request and the response happen behind the scenes. You only deal with typical Java programming language

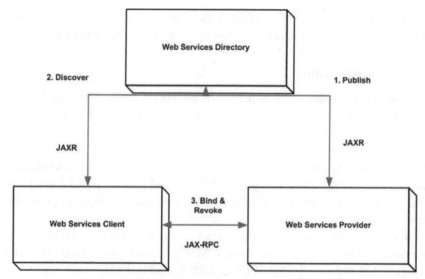

Figure 4.2 Web services Publish-Discover-Invoke model.

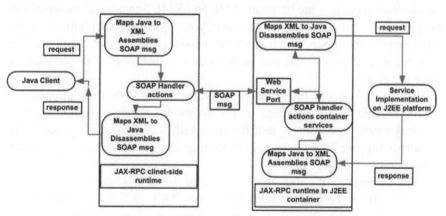

Figure 4.3 A Java client calling a J2EE Web service.

semantics, such as Java method calls, Java data types, and so forth. You need not worry about mapping Java to XML and vice-versa, or constructing SOAP messages. All this low-level work is done behind the scenes, allowing you to focus on the high-level issues.

Note: J2EE 1.4 and Java WSDP 1.5 support both RPC-based and document-oriented Web services. In other words, once a service is discovered, the client can invoke remote procedure calls on the methods offered by the service, or send an XML document to the Web service to be processed.

4.4.9 Interoperability

Interoperability is the most important principle of SOA. This can be realized through the use of Web services, as one of the key benefits of Web services is interoperability, which allows different distributed Web services to run on a variety of software platforms and hardware architectures. The Java programming language is already a champion when it comes to platform independence, and consequently the J2EE 1.4 and Java WSDP 1.5 platforms represent the ideal platforms for developing portable and interoperable Web services.

Interoperability and portability start with the standard specifications themselves. The J2EE 1.4 and Java WSDP 1.5 platforms include the technologies that support SOAP, WSDL, UDDI, and ebXML. This core set of specifications – which are used to describe, publish, enable discovery, and invoke Web services – are based on XML and XML Schema. If you have been keeping up with these core specifications, you know it is difficult to determine which products support which levels (or versions) of the specifications. This task becomes harder when you want to ensure that your Web services are interoperable.

The Web Services Interoperability Organization (WS-I) is an open, industry organization committed to promoting interoperability among Web services based on common, industry-accepted definitions and related XML standards support. WS-I creates guidelines and tools to help developers build interoperable Web services.

WS-I addresses the interoperability need through profiles. The first profile, WS-I Basic Profile 1.0 (which includes XML Schema 1.0, SOAP 1.1, WSDL 1.1, and UDDI 2.0), attempts to improve interoperability within its scope, which is bounded by the specification referenced by it.

Since the J2EE 1.4 and Java WSDP 1.5 platforms adhere to the WS-I Basic Profile 1.0, they ensure not only that applications are portable across J2EE implementations, but also that Web services are interoperable with any Web service implemented on any other platform that conforms to WS-I standards such as .Net [3].

4.4.10 Challenges in Moving to SOA

SOA is usually realized through Web services. Web services specifications may add to the confusion of how to best utilize SOA to solve business problems. In order for a smooth transition to SOA, managers and developers in organizations should known that:

- SOA is an architectural style that has been around for years. Web services are the preferred way to realize SOA.
- SOA is more than just deploying software. Organizations need to analyze their design techniques and development methodology and partner/customer/supplier relationship.
- Moving to SOA should be done incrementally and this requires a shift in how we compose service-based applications while maximizing existing IT investments.

4.5 Summary

This chapter mainly focuses on concepts and role of Business Process Management in Service Oriented Architecture. It also deals with how to work with dynamic Business Process Management and SOA environment. And it also discusses on the co- ordination between Business Process Management, Service Oriented Architecture and Web Services with a short record of its progress.

References

1. Paul C. Brown. BPM in SOA Environment. Tibco Software Inc., http://www.total-architecture.com/indexFiles/Presentations/BPM%20in%20a%20SOA %20Environment.pdf, 2009.
2. PNMSoft. Reducing Information Overhead. http://www.pnmsoft.com/downloads/ Dynamic_BPM.pdf, 2008.
3. Qusay H. Mahmoud. Service-Oriented Architecture (SOA) and Web Services: The Road to Enterprise Application Integration (EAI). Oracle Technology Network, http://www.oracle.com/technetwork/articles/javase/index-142519.html, 2005.
4. Sandeep Arora. Business Process Management. Process Is the Enterprise. SOA Institute, http://www.soainstitute.org/articles/article/article/eai-bpm-and-soa/news-browse/ 1.html, 2005.
5. TIBCO. Dynamic BPM. http://www.fr.tibco.com/software/business-process-management/dynamic-bpm/default.jsp, 2010.

5

Web Service Architecture and Its Specifications

5.1 Introduction

Web services are self-contained, modular applications that can be described, published, located, and invoked over a network. The current trend in the application space is moving away from tightly coupled monolithic systems and towards systems of loosely coupled, dynamically bound components that optimizes agreement and shared context among business systems from different organizations to be reliable for open, low-overhead B2B e-business. Systems built with these principles are more likely to dominate the next generation of e-business systems, with flexibility being the overriding characteristic of their success. Service integration is the innovation of the next generation of e-business taking advantage of e-portals and e-marketplaces and leveraging new technologies on the core.

The concept of Web services is the next generation architecture for e-business that needs multiple implementations. There are many ways to instantiate a Web service by choosing various implementation techniques for the roles, operations, and so on as described by the Web services architecture. The new architectural approach of Web service is different from the monolithic closely coupled architectures and demands more flexible architecture, yielding systems that are more amenable to changes.

5.2 Application Components

5.2.1 Web Service Architecture

Web services demand a new architectural approach that is different from the traditional approach of Web-oriented systems. The traditional systems lack coupling between various components in the system. The traditional Web-oriented systems are monolithic and are more sensitive to changes, therefore any minor change in any of the subsystem has a greater impact on the functioning of the Web-oriented system and hence to build an efficient and effective Web service application current models of application design needs to be replaced with a more flexible architecture, yielding systems that are more amenable to change.

5.2.2 Web Service Components

The Web services architecture is the logical evolution of object-oriented analysis and design, and the logical evolution of components geared towards the architecture, design, implementation, and deployment of e-business solutions. Few fundamental concepts in Web Services are encapsulation, message passing, dynamic binding, service description and querying. Basics of Web services, is the notion that everything is a service, publishing an API for use by other services on the network and encapsulating implementation details.

Web service components should work in co-ordination to perform several essential activities in any service-oriented environment. The component process includes create, publish, locate, invoke and unpublish. The various activities that take place in any service-oriented environment are:

- A Web service needs to be created, and its interfaces and invocation methods must be defined.
- A Web service needs to be published to one or more intranet or Internet repositories for potential users to locate.
- A Web service needs to be located to be invoked by potential users.
- A Web service needs to be invoked to be of any benefit.
- A Web service may need to be unpublished when it is no longer available or needed.

Web services architecture requires three fundamental operations: publish, find, and bind. Service providers publish services to a service broker. Service requesters find required services using a service broker and bind to them. These ideas are shown in Figure 5.1.

The mechanism of service description is one of the key elements in Web service architecture. All collaborations in the Web services architecture have the possibility of being controlled by a configurable, negotiable set of environmental prerequisites. An environmental prerequisite is any nonfunctional component or infrastructure mechanism that must be made operational before a service can be invoked – for example, the use of a particular commu-

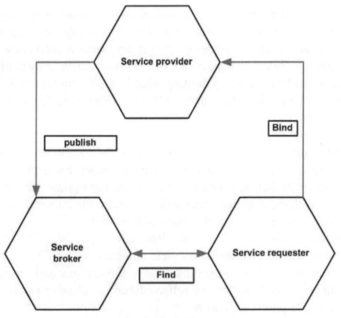

Figure 5.1 Publish, find, and bind.

nications mechanism (HTTPS, IBM MQSeries), or the use of a particular third-party auditing or billing service. These components must be put in place before the service can actually be invoked. A service may support multiple possible implementations for any environmental prerequisite it specifies. For example, the service could offer a choice of communications layer, choice of billing service, or other option. The service requester can then negotiate or choose which implementation to use to satisfy the environmental prerequisite. It is through environmental prerequisites that collaborations can be as secure, reliable, and safe as required by the two collaborators in Web services architecture.

5.2.3 Need for Web Services Architecture

The demanding architecture for Web services yields numerous benefits, a few of which are discussed below:

- *Promotes interoperability by minimizing the requirements for shared understanding*
  XML-based interface definition language (NASSL), an XML-based ser-

vice description (WDS) and a protocol of collaboration and negotiation are the only requirements for shared understanding between a service provider and a service requester. By limiting what is absolutely required for interoperability, collaborating Web services can be truly platform and language independent. By limiting what is absolutely required, Web services can be implemented using a large number of different underlying infrastructures.

- *Enables just-in-time integration*
  Collaborations in Web services are bound dynamically at runtime. A service requester describes the capabilities of the service required and uses the service broker infrastructure to find an appropriate service. Once a service with the required capabilities is found, the information from the service's NASSL document is used to bind to it.
  Dynamic service discovery and invocation (publish, find, bind) and message-oriented collaboration yield applications with looser coupling, enabling just-in-time integration of new applications and services. This in turn yields systems that are self-configuring, adaptive and robust with fewer single points of failure.
- *Reduces complexity by encapsulation*
  All components in Web services are services. What is important is the type of behavior a service provides, not how it is implemented. A WDS document is the mechanism to describe the behavior encapsulated by a service.
  Encapsulation is key to:
    - Coping with complexity. System complexity is reduced when application designers do not have to worry about implementation details of the services they are invoking.
    - Flexibility and scalability. Substitution of different implementation of the same type of service, or multiple equivalent services, is possible at runtime.
    - Extensibility. Behavior is encapsulated and extended by providing new services with similar service descriptions.
- *Enables interoperability of legacy applications*
  By allowing legacy applications to be *wrapped* in NASSL and WDS documents, and exposed as services, the Web services architecture easily enables new interoperability between these applications. In addition, security, middleware and communications technologies can be wrapped

to participate in a Web service as environmental prerequisites. Directory technologies, such as LDAP, can be wrapped to act as a service broker. Through wrapping the underlying plumbing (communications layer, for example), services insulate the application programmer from the lower layers of the programming stack. This allows services to enable virtual enterprises to link their heterogeneous systems as required (through http-based communications) and/or to participate in single, administrative domain situations, where other communications mechanisms (for example, MQSeries) can provide a richer level of functionality. Examples of this can be found in merger situations, where the resulting enterprise must integrate disparate IT systems and business processes. A service-oriented architecture would greatly facilitate a seamless integration between these systems. Another example can be found in the combination of the travel industry with pervasive computing, when largely mainframe-based travel applications can be exposed as services through wrapping and made available for use by various devices in a service-oriented environment. New services can be created and dynamically published and discovered without disrupting the existing environment.

5.3 Elements of Web Services

Web services are application components that communicate with each other using open protocols. Web services are self-contained and self-describing that can be discovered using UDDI and can be used by other applications. XML is the basis for Web services supported by HTML. Therefore the basic Web services platform is a combination of XML + HTTP.

5.3.1 eXtensible Markup Language (XML)

eXtensible Markup Language (XML) is a set of rules for encoding documents in machine-readable form. It is defined in the XML 1.0 Specification produced by the W3C, and several other related specifications. XML's design goals emphasize simplicity, generality, and usability over the Internet. It is a textual data format with strong support via Unicode for the languages of the world. The World Wide Web Consortium (W3C) is an international community where Member organizations, a full-time staff, and the public work together to develop Web standards. Led by Web inventor Tim Berners-Lee and CEO Jeffrey Jaffe, W3C's mission is to lead the Web to its full potential.

W3C's primary activity is to developing protocols and guidelines that ensure long-term growth for the Web. W3C's standards define key parts of what makes the World Wide Web work.

5.3.2 XML and Web Services

XML Web services are the fundamental building blocks in the move to distributed computing on the Internet. Open standards and the focus on communication and collaboration among people and applications have created an environment where XML Web services are becoming the platform for application integration. Applications are constructed using multiple XML Web services from various sources that work together regardless of where they reside or how they were implemented. XML provides a language which can be used between different platforms and programming languages to express complex messages and functions. One of the primary advantages of the XML Web services architecture is that it allows programs written in different languages on different platforms to communicate with each other in a standards-based way. The other significant advantage that XML Web services have over previous applications is that they work with standard Web protocols like XML, HTTP and TCP/IP. The other important XML Web service features are:

- XML Web services expose useful functionality to Web users through a standard Web protocol. In most cases, the protocol used is SOAP.
- XML Web services provide a way to describe their interfaces in enough detail to allow a user to build a client application to talk to them. This description is usually provided in an XML document called a Web services Description Language (WSDL) document.
- XML Web services are registered so that potential users can find them easily. This is done with Universal Discovery Description and Integration (UDDI).

XML Web service can be defined as a software service exposed on the Web through SOAP, described with a WSDL file and registered in UDDI.

5.3.3 HTTP

The HTTP protocol is the most used Internet protocol. The Hypertext Transfer Protocol (HTTP) is a networking protocol for distributed, collaborative, hypermedia information systems. HTTP is the foundation of data communication for the World Wide Web. The HTTP protocol is designed to permit intermediate network elements to improve or enable communications between clients and servers. HTTP is an Application Layer protocol designed within the framework of the Internet Protocol Suite. The protocol definitions presume a reliable Transport Layer protocol for host-to-host data transfer. The Transmission Control Protocol (TCP) is the dominant protocol in use for this purpose. However, HTTP has found application even with unreliable protocols, such as the User Datagram Protocol (UDP) in methods such as the Simple Service Discovery Protocol (SSDP).

5.3.4 Web Services Platform Elements

- SOAP (Simple Object Access Protocol)
- UDDI (Universal Description, Discovery and Integration)
- WSDL (Web services Description Language)

By using Web services, your application can publish its function or message to the rest of the world. Web services use XML to code and to decode data, and SOAP to transport it and it finds two types of uses, which are described below.

5.3.4.1 Reusable Application Components
There are things applications need very often. Web services can offer application components like: currency conversion, weather reports, or even language translation as services.

5.3.4.2 Connect Existing Software
Web services can help to solve the interoperability problem by giving different applications a way to link their data.

With Web services you can exchange data between different applications and different platforms. Web services have three basic platform elements: SOAP, WSDL and UDDI.

SOAP
SOAP stands for Simple Object Access Protocol. SOAP is an XML-based protocol to let applications exchange information over HTTP. In simpler words SOAP is a protocol for accessing a Web service. The basic features of SOAP are as follows:

- SOAP stands for Simple Object Access Protocol and it is mainly designated as a communication protocol.

- SOAP is a format for sending messages and it is designed to communicate via the Internet.
- SOAP is platform independent and language independent and it is based on eXtensible Markup Language (XML) which makes it simple and easy to handle.
- SOAP allows user to get around firewalls and it is a World Wide Consortium (W3C) standard.

WSDL

WSDL stands for Web services Description Language. WSDL is an XML-based language for locating and describing Web services. WSDL is an XML format for describing network services as a set of endpoints operating on messages containing either document-oriented or procedure-oriented information. The basic features of WSDL are as follows:

- WSDL is based on XML.
- WSDL is used to describe Web services.
- WSDL is used to locate Web services.
- WSDL is a W3C standard.

UDDI

UDDI stands for Universal Description, Discovery and Integration. UDDI is a platform-independent framework for describing services, discovering businesses, and integrating business services by using the Internet. UDDI is a directory service where companies can register and search for Web services. The basic features of UDDI are as follows:

- UDDI is a directory for storing information about Web services.
- UDDI is a directory of Web service interfaces described by WSDL.
- UDDI communicates via SOAP.
- UDDI is built into the Microsoft .NET platform.
- UDDI uses World Wide Web Consortium (W3C) and Internet Engineering Task Force (IETF) Internet standards such as XML, HTTP, and DNS protocols.
- UDDI uses WSDL to describe interfaces to Web services [2].

5.4 Web Service Models

Web services provide a standard means of interoperating between different software applications, running on a variety of platforms and/or frameworks.

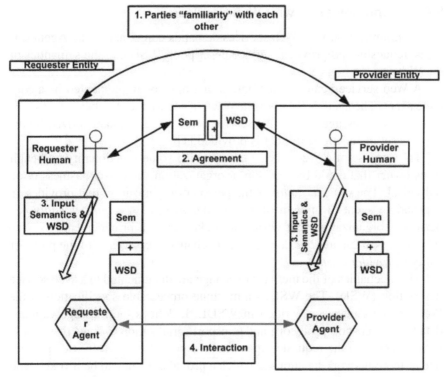

Figure 5.2 The general process of engaging a Web service.

The Web services architecture is interoperability architecture. Web services can be crisply defined as follows:

A Web service is a software system designed to support interoperable machine-to-machine interaction over a network. It has an interface described in a machine-processable format (specifically WSDL). Other systems interact with the Web service in a manner prescribed by its description using SOAP messages, typically conveyed using HTTP with an XML serialization in conjunction with other Web-related standards.

5.4.1 Communication with Web Services

The communication with Web services demands components like Agent services, requester and provider, service description WSD and the semantics of the Web service.

A Web service is an abstract notion that must be implemented by a concrete agent. The agent is the concrete piece of software or hardware that sends and receives messages, while the service is the resource characterized by the abstract set of functionality that is provided [1].

The purpose of a Web service is to provide some functionality on behalf of its owner that could be a person or organization, such as a business or an individual. The *provider entity* is the person or organization that provides an appropriate agent to implement a particular service. A *requester entity* is a person or organization that wishes to make use of a provider entity's Web service. It will use a *requester agent* to exchange messages with the provider entity's *provider agent*.

The mechanics of the message exchange are documented in a Web service description (WSD). The WSD is a machine-processable specification of the Web service's interface, written in WSDL. It defines the message formats, data types, transport protocols, and transport serialization formats that should be used between the requester agent and the provider agent. It also specifies one or more network locations at which a provider agent can be invoked, and may provide some information about the message exchange pattern that is expected. In essence, the service description represents an agreement governing the mechanics of interacting with that service.

The semantics of a Web service is the shared expectation about the behavior of the service, in particular in response to messages that are sent to it. In effect, this is the 'contract' between the requester entity and the provider entity regarding the purpose and consequences of the interaction. Although this contract represents the overall agreement between the requester entity and the provider entity on how and why their respective agents will interact, it is not necessarily written or explicitly negotiated. It may be explicit or implicit, oral or written, machine processable or human-oriented, and it may be a legal agreement or an informal (non-legal) agreement.

While the service description represents a contract governing the mechanics of interacting with a particular service, the semantics represents a contract governing the meaning and purpose of that interaction. The dividing line between these two is not necessarily rigid. As more semantically rich languages are used to describe the mechanics of the interaction, more of the

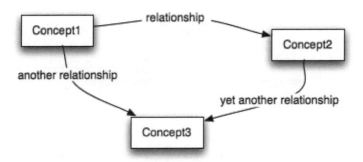

Figure 5.3 Sample concept map.

essential information may migrate from the informal semantics to the service description. As this migration occurs, more of the work required to achieve successful interaction can be automated.

5.4.2 Concept Maps

A concept map is an informal, graphical way to illustrate key concepts and relationships. Each concept is presented in a regular, stylized way consisting of a short definition, an enumeration of the relationships with other concepts, and a slightly longer explanatory description. Relationships denote associations between concepts. Grammatically, relationships are verbs; or more accurately, predicates. A statement of a relationship typically takes the form: concept predicate concept. In a concept map concepts are represented using boxes and relationships using labeled arcs. The merit of a concept map is that it allows rapid navigation of the key concepts and illustrates how they relate to each other. It should be stressed however that these diagrams are primarily navigational aids; the written text is the definitive source.

5.4.3 Service Models

A model is a coherent subset of the architecture that typically revolves around a particular aspect of the overall architecture. Although different models share concepts, it is usually from different points of view; the major role of a model is to explain and encapsulate a significant theme within the overall Web services architecture. The various Web services models are: The Message Oriented Model, The Service-Oriented Model, The Resource Oriented Model, and The Policy Model.

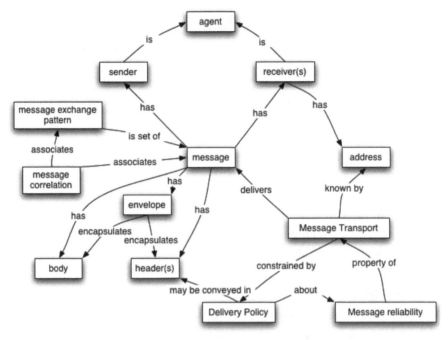

Figure 5.4 Message-oriented model.

5.4.3.1 The Message Oriented Model

The Message Oriented Model focuses on messages, message structure and message transport. The key concepts of this model are: the agent that sends and receives messages, the structure of the message in terms of message headers and bodies and the mechanisms used to deliver messages with additional concepts that includes the role of policies and how they govern the message level model. Specifically, in this model, we are not concerned with any semantic significance of the content of a message or its relationship to other messages.

In this model an *address* is the particular information required by a message transport mechanism in order to deliver a message appropriately. The form of the address information will depend of the particular message transport. In the case of an HTTP message transport, the address information will take the form of a URL.

A *delivery policy* is a policy that constrains the methods by which messages are delivered by the message transport. A delivery policy applies to the combination of a particular message and a particular message transport mechanism. The policies that apply, however, may originate from descriptions in the message itself, or be intrinsic to the transport mechanism, or both [1].

A *message* represents the data structure passed from its sender to its recipients. The structure of a message is defined in a service description. The main parts of a message are its envelope, a set of zero or more headers, and the message body. A message can be as simple as an HTTP GET request or a message can also simply be a plain XML document or SOAP XML document.

The *message body* provides a mechanism for transmitting information to the recipient of the message.

Message correlation is the association of a message with a context. Message correlation ensures that a requester agent can match the reply with the request, especially when multiple replies may be possible.

A *message envelope* is the structure that encapsulates the component parts of a message: the message body and the message headers.

A *Message Exchange Pattern* (MEP) is a template, devoid of application semantics, that describes a generic pattern for the exchange of messages between agents. It describes relationships (e.g., temporal, causal, sequential, etc.) of multiple messages exchanged in conformance with the pattern, as well as the normal and abnormal termination of any message exchange conforming to the pattern.

Message headers represent information about messages that is independently standardized (such as WS-Security) – and may have separate semantics – from the message body.

The *message receiver* is an agent that is intended to receive a message from the message sender. The goal of reliable messaging is to both reduce the error frequency for messaging and to provide sufficient information about the status of a message delivery. Such information enables a participating agent to make a compensating decision when errors or less than desired results occur.

A *message sender* is an agent that transmits a message to another agent. Although every message has a sender, the identity of the sender may not be available to others in the case of anonymous interactions.

A *requester agent* and a *provider agent* exchange a number of messages during an interaction. The ordered set of messages exchanged is a message sequence. This sequence may be realizing a well-defined MEP, usually identified by a URI.

The message transport is the actual mechanism used to deliver messages. For a message transport to function, the sending agent must provide the address of the recipient.

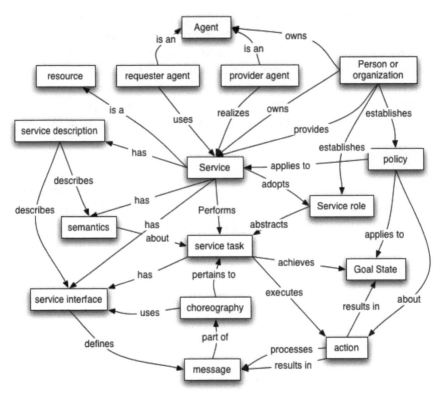

Figure 5.5 Service-Oriented Model (SOM).

5.4.3.2 The Service-Oriented Model

The Service-Oriented Model (SOM) focuses on those aspects of the architecture that relate to service and action. The primary purpose of the SOM is to explicate the relationships between an agent and the services it provides and requests. The SOM builds on the MOM, but its focus is on action rather than message.

An *action*, for the purposes of this architecture, is any action that may be performed by an agent, possibly as a result of receiving a message, or which results in sending a message or another observable state change.

An *agent* is a program acting on behalf of a person or an organization.

Choreography defines the sequence and conditions under which multiple cooperating independent agents exchange messages in order to perform a task to achieve a goal state. Choreography can be distinguished from an orchestration. An orchestration defines the sequence and conditions in which one Web service invokes other Web services in order to realize some useful function.

Capability is a named piece of functionality or feature that is declared as supported or requested by an agent.

Goal states are associated with tasks. Tasks are the unit of action associated with services that have a measurable meaning.

The *provider agent* is the software agent that realizes a Web service by performing tasks on behalf of its owner.

A *requester agent* is a software agent that wishes to interact with a provider agent in order to request that a task be performed on behalf of its owner.

The *requester entity* is the person or organization that wishes to use a provider entity's Web service.

Service is an abstract resource that represents a capability of performing tasks that represents a coherent functionality from the point of view of provider entities and requester entities.

Service description is a set of documents that describe the interface to and semantics of a service.

Service interface is the abstract boundary that a service exposes. It defines the types of messages and the message exchange patterns that are involved in interacting with the service, together with any conditions implied by those messages. A service interface defines the different types of messages that a service sends and receives, along with the message exchange patterns that may be used.

Service intermediary is a Web service whose main role is to transform messages in a value-added way starting from a messaging point of view, an intermediary processes messages en route from one agent to another.

Service role is an abstract set of tasks which is identified to be relevant by a person or organization offering a service. Service roles are also associated with particular aspects of messages exchanged with a service.

The *semantics* of a service is the behavior expected when interacting with the service. The semantics expresses a contract between the provider entity and the requester entity.

A *service task* is an action or combination of actions that is associated with a desired goal state. Performing the task involves executing the actions, and is intended to achieve a particular goal state [1].

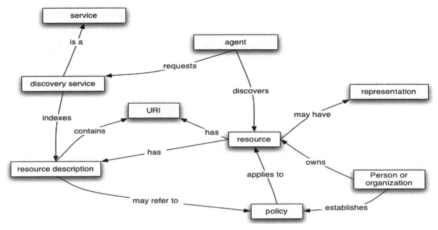

Figure 5.6 Resource-Oriented Model.

5.4.3.3 The Resource-Oriented Model

The Resource-Oriented Model focuses on those aspects of the architecture that relate to resources. Resources are a fundamental concept that underpins much of the Web and much of Web services. The ROM focuses on the key features of resources that are relevant to the concept of resource, independent of the role the resource has in the context of Web services.

Discovery is the act of locating a machine-processable description of a Web service-related resource that may have been previously unknown and that meets certain functional criteria. It involves matching a set of functional and other criteria with a set of resource descriptions.

A *discovery service* is a service that enables agents to retrieve Web service-related resource descriptions.

Identifiers are used to identify resources. In the architecture we use Uniform Resource Identifiers [RFC 2396] to identify resources.

Representations are data objects that reflect the state of a resource.

A *resource* is defined by [RFC 2396] to be anything that can have an identifier. Resources form the heart of the Web architecture itself.

A *resource description* is any machine readable data that may permit resources to be discovered. Resource descriptions may be of many different forms, tailored for specific purposes, but all resource descriptions must contain the resource's identifier.

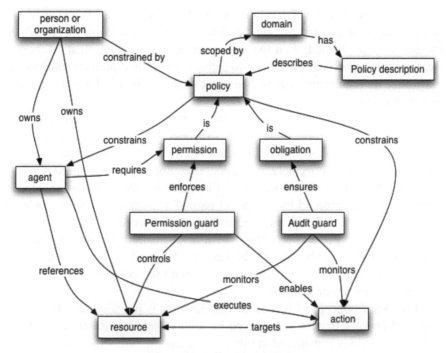

Figure 5.7 Policy Model (PM).

5.4.3.4 The Policy Model

The Policy Model (PM) focuses on those aspects of the architecture that relate to policies and, by extension, security and quality of service. Security is fundamentally about constraints; about constraints on the behavior on action and on accessing resources. Similarly, quality of service is also about constraints on service. In the PM, these constraints are modeled around the core concept of policy; and the relationships with other elements of the architecture. Thus the PM is a framework in which security can be realized.

An *audit guard* is a mechanism used on behalf of an owner that monitors actions and agents to verify the satisfaction of obligations. Typically, an audit guard monitors the state of a resource or a service, ensuring that the obligation is satisfied. It determines whether the associated obligations are satisfied.

A *domain* defines the scope of applicability of policies. A domain may be defined explicitly or implicitly. Members of an explicitly defined domain are enumerated by a central authority; members of an implicitly defined domain are not.

An *obligation* is one of two fundamental types of policies. When an agent has an obligation to perform some action, then it is required to do so. When the action is performed, then the agent can be said to have satisfied its obligations. Not all obligations relate to actions. An obligation may continue to exist after its requirements have been met or it may be discharged by some action or event.

Permission is one of two fundamental types of policies. When an agent has permission to perform some action, to access some resource, or to achieve a certain state, then it is expected that any attempt to perform the action, etc., will be successful. Conversely, if an agent does not have the required permission, then the action should fail even if it would otherwise have succeeded. Permissions are enforced by guards, in particular permission guards, whose function is to ensure that permission policies are honored.

A *permission guard* is an enforcement mechanism that is used to enforce permission policies. The role of the permission guard is to ensure that any uses of a service or resource are consistent with the policies established by the service's owner or manager.

The WSA concept of person or organization is intended to refer to the real-world people that are represented by agents that perform actions on their behalf. All actions considered in this architecture are ultimately rooted in the actions of humans.

A *policy* is a constraint on the behavior of agents as they perform actions or access resources. There are many kinds of policies, some relate to accessing resources in particular ways, others relate more generally to the allowable actions an agent may perform: both as provider agents and as requester agents.

A *policy description* is a machine processable description of some constraint on the behavior of agents as they perform actions, access resources.

A *policy guard* is an abstraction that denotes a mechanism that is used by owners of resources to enforce policies.

5.5 REST Architecture

REST is a term coined by Roy Fielding to describe an architecture style of networked systems. REST is an acronym standing for Representational State Transfer.

> Representational State Transfer is intended to evoke an image of how a well-designed Web application behaves: a network of Web

pages (a virtual state-machine), where the user progresses through an application by selecting links (state transitions), resulting in the next page (representing the next state of the application) being transferred to the user and rendered for their use.

REST is a way to access documents and resources using simple remote software architecture, represented by an API. The Web (World Wide Web) is an application of the REST architecture. The REST architecture does not necessarily use HTTP or the Web. This architecture is defined as follows:

- States and functions of a remote application are considered as resources.
- Each resource is only accessible using a standard address called hyperlinks or URIs. Resources have a standard interface in which the transactions and data types are precisely defined.

The motivation for REST was to capture the characteristics of the Web which made the Web successful. REST is not a standard. REST is just an architectural style but it uses other standards like [2]:

- HTTP,
- URL,
- XML/HTML/GIF/JPEG, etc. (Resource Representations),
- text/xml, text/html, image/gif, image/jpeg, etc. (MIME types).

5.5.1 REST Web Services Characteristics

The characteristics of REST architecture are:

- Client-Server: a pull-based interaction style, consuming components pull representations.
- Stateless: each request from client to server must contain all the information necessary to understand the request, and cannot take advantage of any stored context on the server.
- Cache: to improve network efficiency responses must be capable of being labeled as cacheable or non-cacheable.
- Uniform interface: all resources are accessed with a generic interface (e.g., HTTP GET, POST, PUT, DELETE).
- Named resources: the system is comprised of resources which are named using a URL.
- Interconnected resource representations: the representations of the resources are interconnected using URLs, thereby enabling a client to progress from one state to another.

- Layered components: intermediaries, such as proxy servers, cache servers, gateways, etc., can be inserted between clients and resources to support performance, security, etc.

5.5.2 Principles of REST Web Service Design

- The key to creating Web services in a REST network (i.e., the Web) is to identify all of the conceptual entities that user wish to expose as services.
- Create a URL to each resource. The resources should be nouns, not verbs.
- Bottom of Form.
- Categorize resources according to whether clients can just receive a representation of the resource, or whether clients can modify (add to) the resource. For the former, make those resources accessible using an HTTP GET. For the later, make those resources accessible using HTTP POST, PUT, and/or DELETE.
- All resources accessible via HTTP GET should be side-effect free. That is, the resource should just return a representation of the resource. Invoking the resource should not result in modifying the resource.
- No representation should be an island. In other words, put hyperlinks within resource representations to enable clients to drill down for more information, and/or to obtain related information.
- Design to reveal data gradually. Do not reveal everything in a single response document. Provide hyperlinks to obtain more details.
- Specify the format of response data using a schema (DTD, W3C Schema, RelaxNG, or Schematron). For those services that require a POST or PUT to it, also provide a schema to specify the format of the response.
- Describe how your services are to be invoked using either a WSDL document, or simply an HTML document.

5.5.3 REST Architecture

REST-style architectures consist of clients and servers. Clients initiate requests to servers; servers process requests and return appropriate responses. Requests and responses are built around the transfer of 'representations' of 'resources'. A resource can be essentially any coherent and meaningful concept that may be addressed. A representation of a resource is typically a document that captures the current or intended state of a resource.

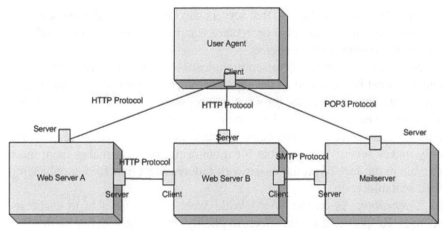

Figure 5.8 Overview of REST architecture.

At any particular time, a client can either be in transition between application states or 'at rest'. A client in a rest state is able to interact with its user, but creates no load and consumes no per-client storage on the set of servers or on the network.

The client begins sending requests when it is ready to make the transition to a new state. While one or more requests are outstanding, the client is considered to be in transition. The representation of each application state contains links that may be used next time the client chooses to initiate a new state transition.

REST was initially described in the context of HTTP, but is not limited to that protocol. RESTful architectures can be based on other Application Layer protocols if they already provide a rich and uniform vocabulary for applications based on the transfer of meaningful representational state. RESTful applications maximize the use of the pre-existing, well-defined interface and other built-in capabilities provided by the chosen network protocol, and minimize the addition of new application-specific features on top of it.

The REST architectural style describes the following six constraints applied to the architecture, while leaving the implementation of the individual components free to design [3].

Client-Server: Clients are separated from servers by a uniform interface. This separation of concerns means that, for example, clients are not concerned with data storage, which remains internal to each server, so that the portability of client code is improved. Servers are not concerned with the user interface or user state, so that servers can be simpler and more scalable. Servers and clients may also be replaced and developed independently, as long as the interface is not altered.

Stateless: The client-server communication is further constrained by no client context being stored on the server between requests. Each request from any client contains all of the information necessary to service the request, and any state is held in the client. The server can be stateful, this constraint merely requires that server-side state be addressable by URL as a resource. This not only makes servers more visible for monitoring, but also makes them more reliable in the face of partial or network failures as well as further enhancing their scalability.

Cacheable: As on the World Wide Web, clients are able to cache responses. Responses must therefore, implicitly or explicitly, define themselves as cacheable or not to prevent clients reusing stale or inappropriate data in response to further requests. Well-managed caching partially or completely eliminates some client-server interactions, further improving scalability and performance.

Layered system: A client cannot ordinarily tell whether it is connected directly to the end server, or to an intermediary along the way. Intermediary servers may improve system scalability by enabling load balancing and by providing shared caches. They may also enforce security policies.

Code on demand (optional)

- Servers are able temporarily to extend or customize the functionality of a client by transferring logic to it that it can execute. Examples of this may include compiled components such as Java applets and client-side scripts such as JavaScript.

Uniform interface

- The uniform interface between clients and servers, discussed below, simplifies and decouples the architecture, which enables each part to evolve independently. The four guiding principles of this interface are detailed below.

The only optional constraint of REST architecture is code on demand. If a service violates any other constraint, it cannot strictly be referred to as RESTful.

Complying with these constraints, and thus conforming to the REST architectural style, will enable any kind of distributed hypermedia system to have desirable emergent properties, such as performance, scalability, simplicity, modifiability, visibility, portability and reliability.

5.5.4 Benefits of REST Architecture

- Easier to implement than the alternatives.
- Just a browser to access a service.
- Caching of resources, thus speeding up operations.
- Less memory consumption than others.
- Possibility to distribute queries across multiple servers.
- Using standard formats as HTML or XML ensures compatibility over time.
- We can exchange requests between various media applications as they are represented by URIs.

5.6 Summary

This chapter highlights on what a Web service is all about and the different architectures and models that support efficient usage of Web based applications and Web services and it also discusses on the REST architecture and its implementation and working along with potential benefits of the architecture with a short history of its development.

References

1. David Booth, Francis McCabe, Eric Newcomer, Iona Michael Champion, Chris Ferris, and David Orchard. Web Services Architecture. http://www.w3.org/TR/ws-arch/, 2004.
2. IBM. Web Services Architecture Overview. http://www.ibm.com/developerworks/library/w-ovr/, 2000.
4. Roger L. Costello. Building Web Services the REST Way. XFront Tutorial http://www.xfront.com/REST-Web-Services.html, 2010.
5. Roy Thomas Fielding. Architectural Styles and the Design of Network-based Software Architectures. http://www.ics.uci.edu/~fielding/pubs/dissertation/rest_arch_style.htm, 2000.
6. scriptol.com. REST – Representational State Transfer. http://www.scriptol.com/programming/rest.php, 2008.

6.6 Summary

This chapter presented...

References

6

Web Service Protocols and Technologies

6.1 eXtensible Markup Language (XML)

eXtensible Markup Language (XML) is a set of rules for encoding documents in machine-readable form. It is defined in the XML 1.0 Specification produced by the W3C, and several other related specifications.

XML's design goals emphasize simplicity, generality, and usability over the Internet. It is a textual data format with strong support via Unicode for the languages of the world. Although the design of XML focuses on documents, it is widely used for the representation of arbitrary data structures, for example in Web services.

Many application programming interfaces (APIs) have been developed that software developers use to process XML data, and several schema systems exist to aid in the definition of XML-based languages.

At present, hundreds of XML-based languages have been developed including RSS, Atom, SOAP, and XHTML. XML-based formats have become the default for most office-productivity tools, including Microsoft Office (Office Open XML), OpenOffice.org (Open Document), and Apple's iWork.

6.1.1 What Is XML?

- XML stands for EXtensible Markup Language.
- XML is a markup language much like HTML.
- XML was designed to carry data, not to display data.
- XML tags are not predefined. You must define your own tags.
- XML is designed to be self-descriptive.
- XML is a W3C Recommendation.

6.1.2 History of XML

XML is an application profile of SGML (ISO 8879). The versatility of SGML for dynamic information display was understood by early digital media publishers in the late 1980s prior to the rise of the Internet. By the mid-1990s some practitioners of SGML had gained experience with the then-new World Wide Web, and believed that SGML offered solutions to some of the problems the Web was likely to face as it grew. Dan Connolly added SGML to the list of W3C's activities when he joined the staff in 1995, the work began in mid-1996 when Sun Microsystems engineer Jon Bosak developed a charter and recruited collaborators. Bosak was well connected in the small community of people who had experience both in SGML and the Web.

XML was compiled by a working group of eleven members, supported by an (approximately) 150-member Interest Group. Technical debate took place on the Interest Group mailing list and issues were resolved by consensus or, when that failed, majority vote of the Working Group. A record of design decisions and their rationales was compiled by Michael Sperberg-McQueen on December 4, 1997. James Clark served as Technical Lead of the Working Group, notably contributing the empty-element '<empty/>' syntax and the name 'XML'. Other names that had been put forward for consideration included 'MAGMA' (Minimal Architecture for Generalized Markup Applications), 'SLIM' (Structured Language for Internet Markup) and 'MGML' (Minimal Generalized Markup Language). The co-editors of the specification were originally Tim Bray and Michael Sperberg-McQueen.

The XML Working Group never met face-to-face; the design was accomplished using a combination of email and weekly teleconferences. The major design decisions were reached in twenty weeks of intense work between July and November 1996, when the first Working Draft of an XML specification was published. Further design work continued through 1997, and XML 1.0 became a W3C Recommendation on February 10, 1998.

6.1.3 Key Terminology

The key terminologies that are associated with XML and also used in regular XML applications are:

(Unicode) Character

By definition, an XML document is a string of characters. Almost every legal Unicode character may appear in an XML document.

Processor and Application

The processor analyzes the markup and passes structured information to an application. The specification places requirements on what an XML processor must do and not do. The processor is often referred to colloquially as an XML parser.

Markup and Content

The characters which make up an XML document are divided into markup and content. Markup and content may be distinguished by the application of simple syntactic rules. All strings which constitute markup either begin with the character '<' and end with a '>', or begin with the character '&' and end with a ';'. Strings of characters which are not markup are content.

Tag

A markup construct that begins with '<' and ends with '>'. Tags come in three flavors: start-tags, for example `<section>`, end-tags, for example `</section>`, and empty-element tags, for example `<line-break/>`.

Element

A logical component of a document which either begins with a start-tag and ends with a matching end-tag, or consists only of an empty-element tag. The characters between the start and end-tags, if any, are the element's content, and may contain markup, including other elements, which are called child elements. An example of an element is `<Greeting>Hello, world.</Greeting>` (see hello world). Another is `<line-break/>`.

Attribute

A markup construct consisting of a name/value pair that exists within a start-tag or empty-element tag. In the example the element img has two attributes, src and alt: `<img src="scenery.jpg" alt="by Brooks"/>`. Another example would be `<step number="3">Connect A to B.</step>` where the name of the attribute is 'number' and the value is '3'.

6.1.4 The Difference between XML and HTML

XML is not a replacement for HTML. XML and HTML were designed with different goals:

- XML was designed to transport and store data, with focus on what data is.

- HTML was designed to display data, with focus on how data looks.
- HTML is about displaying information, while XML is about carrying information.

6.1.5 Dormancy of XML

XML was created to structure, store, and transport information only and not to go on with any additional processing. The following example is a note to Tove, from Jani, stored as XML:

```
<note>
<to>Tove</to>
<from>Jani</from>
<heading>Reminder</heading>
<body>Don't forget me this weekend!</body>
</note>
```

The note above is quite self descriptive. It has sender and receiver information, it also has a heading and a message body. But still, this XML document does not DO anything. It is just information wrapped in tags.

6.1.6 Features of XML

6.1.6.1 XML as Data Type

XML is beginning to appear as a first-class data type in other languages. The ECMAScript for XML (E4X) extension to the ECMAScript/JavaScript language explicitly defines two specific objects (XML and XMLList) for JavaScript, which support XML document nodes and XML node lists as distinct objects and use dot-notation specifying parent-child relationships. E4X is supported by the Mozilla 2.5+ browsers and Adobe Action script, but has not been adopted more universally. Similar notations are used in Microsoft's LINQ implementation for Microsoft .NET 3.5 and above, and in Scala (which uses the Java VM). The open-source xmlsh application, which provides a Linux-like shell with special features for XML manipulation, similarly treats XML as a data type, using the <[ ]> notation. The Resource Description Framework defines a data type RDF: XMLLiteral to hold wrapped, canonical XML.

6.1.6.2 XML Separates Data from HTML

To display dynamic data in your HTML document, it will take a lot of work to edit the HTML each time the data changes. With XML, data can be stored in separate XML files. This way it is possible to concentrate on using HTML for layout and display, and be sure that changes in the underlying data will not require any changes to the HTML. With few lines of JavaScript code, you can read an external XML file and update the data content of your Web page.

6.1.6.3 XML Simplifies Data Sharing

In the real world, computer systems and databases contain data in incompatible formats. XML data is stored in plain text format. This provides a software- and hardware-independent way of storing data. This makes it much easier to create data that can be shared by different applications.

6.1.6.4 XML Simplifies Data Transport

One of the most time-consuming challenges for developers is to exchange data between incompatible systems over the Internet. Exchanging data as XML greatly reduces this complexity, since the data can be read by different incompatible applications.

6.1.6.5 XML Simplifies Platform Changes

Upgrading to new systems is always time consuming. Large amounts of data must be converted and incompatible data is often lost. XML data is stored in text format. This makes it easier to expand or upgrade to new operating systems, new applications, or new browsers, without losing data.

6.1.6.6 XML Makes Your Data More Available

Different applications can access data, not only in HTML pages, but also from XML data sources. With XML, data can be available to all kinds of 'reading machines' (handheld computers, voice machines, news feeds, etc.), and make it more available for blind people, or people with other disabilities.

6.1.6.7 With XML You Invent Your Own Tags

The tags in the example above (like <to> and <from>) are not defined in any XML standard. These tags are 'invented' by the author of the XML document. That is because the XML language has no predefined tags. The tags used in HTML are predefined. HTML documents can only use tags defined in the HTML standard (like <p>, <h1>, etc.). XML allows the author to define his/her own tags and his/her own document structure.

6.1.6.8 XML Is Used to Create New Internet Languages

A lot of new Internet languages are created with XML. Here are some examples:

- XHTML.
- WSDL for describing available Web services.
- WAP and WML as markup languages for handheld devices.
- RSS languages for news feeds.
- RDF and OWL for describing resources and ontology.
- SMIL for describing multimedia for the Web.

6.1.7 XML Syntax Rules

The syntax rules of XML are very simple and logical. The rules are easy to learn, and easy to use.

6.1.7.1 All XML Elements Must Have a Closing Tag

In HTML, elements do not have to have a closing tag:

```
<p>This is a paragraph
<p>This is another paragraph
```

In XML, it is illegal to omit the closing tag. All elements must have a closing tag:

```
<p>This is a paragraph</p>
<p>This is another paragraph</p>
```

6.1.7.2 XML Tags Are Case Sensitive

XML tags are case sensitive. The tag <Letter> is different from the tag <letter>. Opening and closing tags must be written with the same case:

```
<Message>This is incorrect</message>
<message>This is correct</message>
```

Note: 'Opening and closing tags' are often referred to as 'start and end tags'. Use whatever you prefer. It is exactly the same thing.

6.1.7.3 XML Elements Must Be Properly Nested

In HTML, you might see improperly nested elements:

```
<b><i>This text is bold and italic</b></i>
```

In XML, all elements must be properly nested within each other:

```
<b><i>This text is bold and italic</i></b>
```

In the example above, 'properly nested' simply means that since the `<i>` element is opened inside the `<b>` element, it must also be closed inside the `<b>` element.

6.1.7.4 XML Documents Must Have a Root Element

XML documents must contain one element that is the parent of all other elements. This element is called the root element:

```
<root>
  <child>
    <subchild>.....</subchild>
  </child>
</root>
```

6.1.7.5 XML Attribute Values Must Be quoted

XML elements can have attributes in name/value pairs just like in HTML. In XML, the attribute values must always be quoted. Study the two XML documents below. The first one is incorrect, the second is correct:

```
<note date=12/11/2007>
  <to>Tove</to>
  <from>Jani</from>
</note>
```

```
<note date="12/11/2007">
  <to>Tove</to>
  <from>Jani</from>
</note>
```

The error in the first document is that the date attribute in the note element is not quoted.

6.1.7.6 Entity References

Some characters have a special meaning in XML. If you place a character like '<' inside an XML element, it will generate an error because the parser interprets it as the start of a new element. This will generate an XML error:

```
<message>if salary < 1000 then</message>
```

To avoid this error, replace the '<' character with an entity reference:

```
<message>if salary &lt; 1000 then</message>
```

There are five predefined entity references in XML:

<	<	less than
>	>	greater than
&	&	ampersand
'	'	apostrophe
"	"	quotation mark

Note: Only the characters '<', and '&' are strictly illegal in XML. The 'greater than' character is legal, but it is a good habit to replace it.

6.1.7.7 Comments in XML
The syntax for writing comments in XML is similar to that of HTML.

```
<!-- This is a comment -->
```

6.1.7.8 White-Space Is Preserved in XML
HTML truncates multiple white-space characters to one single white-space:

HTML: Hello   Tove
Output: Hello Tove

With XML, the white-space in a document is not truncated.

6.1.7.9 XML Stores New Line as LF
In Windows applications, a new line is normally stored as a pair of characters: carriage return (CR) and line feed (LF). In Unix applications, a new line is normally stored as a LF character. Macintosh applications also use an LF to store a new line. XML stores a new line as LF.

6.1.7.10 XML Declaration
XML documents may begin by declaring some information about themselves, as in the following example:

```
<?xml version="1.0" encoding="UTF-8" ?>
```

Example
Here is a small, complete XML document, which uses all of these constructs and concepts.

```
<?xml version="1.0" encoding="UTF-8" ?>
<Painting>
  <img src="madonna.jpg" alt='Foligno Madonna, by Raphael'/>
  <caption>This is Raphael's "Foligno" Madonna, painted in
    <date>1511</date>?<date>1512</date>.
  </caption>
</painting>
```

There are five elements in this example document: painting, img, caption, and two dates. The date elements are children of caption, which is a child of the root element painting. img has two attributes, src and alt.

XML Summary

- XML can be used to exchange, share, and store data.
- XML documents form a tree structure that starts at 'the root' and branches to 'the leaves'.
- XML has very simple syntax rules. XML with correct syntax is 'Well Formed'. Valid XML also validates against a DTD.
- XSLT is used to transform XML into other formats like HTML.
- All modern browsers have a built-in XML parser that can read and manipulate XML.
- The DOM (Document Object Model) defines a standard way for accessing XML.
- The XMLHttpRequest object provides a way to communicate with a server after a Web page has loaded.
- XML Namespaces provide a method to avoid element name conflicts.
- Text inside a CDATA section is ignored by the parser.

6.2 Simple Object Access Protocol (SOAP)

SOAP, originally defined as Simple Object Access Protocol, is a protocol specification for exchanging structured information in the implementation of Web Services in computer networks. It relies on XML for its message format, and usually relies on other Application Layer protocols, most notably Remote Procedure Call (RPC) and Hypertext Transfer Protocol (HTTP), for message negotiation and transmission. SOAP can form the foundation layer of a Web services protocol stack, providing a basic messaging framework upon which Web services can be built. This XML-based protocol consists of three parts: an envelope, which defines what is in the message and how to process it, a set of encoding rules for expressing instances of application-defined data types, and a convention for representing procedure calls and responses.

As a layman's example of how SOAP procedures can be used, a SOAP message could be sent to a Web-service-enabled Web site, for example, a real-estate price database, with the parameters needed for a search. The site would then return an XML-formatted document with the resulting data, e.g., prices, location, features. Because the data is returned in a standardized machine-parseable format, it could then be integrated directly into a third-party Web site or application.

The SOAP architecture consists of several layers of specifications: for message format, Message Exchange Patterns (MEP), underlying transport protocol bindings, message processing models, and protocol extensibility. SOAP is the successor of XML-RPC.

6.2.1 What Is SOAP?

- SOAP stands for Simple Object Access Protocol.
- SOAP is a communication protocol.
- SOAP is for communication between applications.
- SOAP is a format for sending messages.
- SOAP is designed to communicate via Internet.
- SOAP is platform independent.
- SOAP is language independent.
- SOAP is based on XML.
- SOAP is simple and extensible.
- SOAP allows you to get around firewalls.
- SOAP will be developed as a W3C standard.

6.2.2 History

SOAP once stood for 'Simple Object Access Protocol' but this acronym was dropped with Version 1.2 of the standard. Version 1.2 became a W3C recommendation on June 24, 2003. The acronym is sometimes confused with SOA, which stands for Service-Oriented Architecture; however SOAP is different from SOA.

SOAP was originally designed by Dave Winer, Don Box, Bob Atkinson, and Mohsen Al-Ghosein in 1998 in a project for Microsoft (where Atkinson and Al-Ghosein were already working at the time) [2], as an object-access protocol. The SOAP specification is currently maintained by the XML Protocol Working Group of the World Wide Web Consortium.

After SOAP was first introduced, it became the underlying layer of a more complex set of Web Services, based on Web Services Description Language (WSDL) and Universal Description Discovery and Integration (UDDI). These services, especially UDDI, have proved to be of far less interest, but an appreciation of them gives a fuller understanding of the expected role of SOAP compared to how Web services have actually developed.

6.2.3 Difference between SOAP and Other RPCs

There are other Remote Procedure Calls (RPCs) style protocols that help in communication between applications like Distributed Component Object Model (DCOM), IIOP (Internet Inter Orb Protocol) and RMI (Remote Method Invocations). The advantages of Soap over these protocols are that Soap uses XML and is text-based, whereas the others are dependent on object-model-specific protocols. Moreover they are not adaptable to the internet whereas Soap uses HTTP protocol.

6.2.4 Usage of SOAP

One of the most important uses of SOAP is to help enable XML Web Services. A Web Service is an application provided as a service on the Web. They are functional software components that can be accessed over the Internet. Web Services combines the best of component-based development and are based on Internet Standards that supports communication over the net.

There are many possible applications for SOAP, including:

- *Business to Business integration* – SOAP allows businesses to develop their applications, and then make those applications available to other companies.
- *Distributed applications* – programs like databases could be stored on one server and accessed and managed by clients across the Internet.

6.2.5 The SOAP Specification

The SOAP specification defines the messaging framework which consists of:

- The SOAP processing model defining the rules for processing a SOAP message.

- The SOAP extensibility model defining the concepts of SOAP features and SOAP modules.
- The SOAP underlying protocol binding framework describing the rules for defining a binding to an underlying protocol that can be used for exchanging SOAP messages between SOAP nodes.
- The SOAP message construct defining the structure of a SOAP message.

6.2.6 SOAP Processing Model

The SOAP processing model describes a distributed processing model, its participants, the SOAP nodes and how a SOAP receiver processes a SOAP message. The following SOAP nodes are defined:

- *SOAP sender* – A SOAP node that transmits a SOAP message.
- *SOAP receiver* – A SOAP node that accepts a SOAP message.
- *SOAP message path* – The set of SOAP nodes through which a single SOAP message passes.
- *Initial SOAP sender (Originator)* – The SOAP sender that originates a SOAP message at the starting point of a SOAP message path.
- *SOAP intermediary* – A SOAP intermediary is both a SOAP receiver and a SOAP sender and is targetable from within a SOAP message. It processes the SOAP header blocks targeted at it and acts to forward a SOAP message towards an ultimate SOAP receiver.
- *Ultimate SOAP receiver* – The SOAP receiver that is a final destination of a SOAP message. It is responsible for processing the contents of the SOAP body and any SOAP header blocks targeted at it. In some circumstances, a SOAP message might not reach an ultimate SOAP receiver, for example because of a problem at a SOAP intermediary. An ultimate SOAP receiver cannot also be a SOAP intermediary for the same SOAP message.

6.2.7 SOAP Building Blocks

A SOAP message is an ordinary XML document containing the following elements:

- An Envelope element that identifies the XML document as a SOAP message.
- A Header element that contains header information.
- A Body element that contains call and response information.
- A Fault element containing errors and status information.

Syntax Rules

Here are some important syntax rules:

- A SOAP message MUST be encoded using XML.
- A SOAP message MUST use the SOAP Envelope namespace.
- A SOAP message MUST use the SOAP Encoding namespace.
- A SOAP message must NOT contain a DTD reference.
- A SOAP message must NOT contain XML Processing Instructions [1].

Skeleton SOAP Message

```
<?xml version="1.0"?>
<soap:Envelope
xmlns:soap="http://www.w3.org/2001/12/soap-envelope"
soap:encodingStyle="http://www.w3.org/2001/12/soap-encoding">
<soap:Header>
...
</soap:Header>

<soap:Body>
...
  <soap:Fault>
  ...
  </soap:Fault>
</soap:Body>

</soap:Envelope>
```

6.2.8 Transport Methods

SOAP makes use of an internet application layer protocol as a transport protocol. Critics have argued that this is an abuse of such protocols, as it is not their intended function and therefore not a role they fulfill well. Proponents of SOAP have drawn analogies to successful uses of protocols at various levels for tunneling other protocols.

Both SMTP and HTTP are valid application layer protocols used as Transport for SOAP, but HTTP has gained wider acceptance as it works well with today's Internet infrastructure; specifically, HTTP works well with network firewalls. SOAP may also be used over HTTPS (which is the same protocol as HTTP at the application level, but uses an encrypted transport protocol underneath) with either simple or mutual authentication; this is the advocated WS-I method to provide Web service security as stated in the WS-

I Basic Profile. This is a major advantage over other distributed protocols like General Inter-ORB Protocol/Internet Inter-Orb Protocol (GIOP/IIOP) or Distributed Component Object Model (DCOM) which are normally filtered by firewalls

6.2.9 Message Format

XML was chosen as the standard message format because of its widespread use by major corporations and open source development efforts. Additionally, a wide variety of freely available tools significantly eases the transition to a SOAP-based implementation. The somewhat lengthy syntax of XML can be both a benefit and a drawback. While it promotes readability for humans, facilitates error detection, and avoids interoperability problems such as byte-order, it can retard processing speed and can be cumbersome. For example, CORBA, GIOP, and DCOM use much shorter, binary message formats. On the other hand, hardware appliances are available to accelerate processing of XML messages. Binary XML is also being explored as a means for streamlining the throughput requirements of XML.

Sample SOAP message

POST /InStock HTTP/1.1
Host: www.example.org
Content-Type: application/soap+xml; charset=utf-8
Content-Length: nnn

```
<?xml version="1.0"?>
<soap:Envelope
xmlns:soap="http://www.w3.org/2001/12/soap-envelope"
soap:encodingStyle="http://www.w3.org/2001/12/soap-encoding">

<soap:Body xmlns:m="http://www.example.org/stock">
  <m:GetStockPrice>
    <m:StockName>IBM</m:StockName>
  </m:GetStockPrice>
</soap:Body>

</soap:Envelope>
```

SOAP (Simple Object Access Protocol) is a way for a program running in one kind of operating system (such as Windows 2000) to communicate with a program in the same or another kind of an operating system (such as Linux)

by using the World Wide Web's Hypertext Transfer Protocol (HTTP) and its eXtensible Markup Language (XML) as the mechanisms for information exchange. Since Web protocols are installed and available for use by all major operating system platforms, HTTP and XML provide an already at-hand solution to the problem of how programs running under different operating systems in a network can communicate with each other. SOAP specifies exactly how to encode an HTTP header and an XML file so that a program in one computer can call a program in another computer and pass it information. It also specifies how the called program can return a response.

SOAP was developed by Microsoft, Develop Mentor, and User land Software and has been proposed as a standard interface to the Internet Engineering Task Force (IETF). It is somewhat similar to the Internet Inter-ORB Protocol (IIOP), a protocol that is part of the Common Object Request Broker Architecture (CORBA). Sun Microsystems' Remote Method Invocation (RMI) is a similar client/server interprogram protocol between programs written in Java.

An advantage of SOAP is that program calls are much more likely to get through firewall servers that screen out requests other than those for known applications (through the designated port mechanism). Since HTTP requests are usually allowed through firewalls, programs using SOAP to communicate can be sure that they can communicate with programs anywhere [1].

6.3 Web Services Description Language (WSDL)

The Web Services Description Language (WSDL, pronounced 'wiz-del') is an XML-based language that provides a model for describing Web services. WSDL is an XML format for describing network services as a set of endpoints operating on messages containing either document-oriented or procedure-oriented information. The operations and messages are described abstractly, and then bound to a concrete network protocol and message format to define an endpoint. Related concrete endpoints are combined into abstract endpoints (services). WSDL is extensible to allow description of endpoints and their messages regardless of what message formats or network protocols are used to communicate, however, the only bindings described in this document describe how to use WSDL in conjunction with SOAP 1.1, HTTP GET/POST, and MIME.

6.3.1 What Is WSDL?

- WSDL stands for Web Services Description Language.
- WSDL is written in XML.
- WSDL is an XML document.
- WSDL is used to describe Web services.
- WSDL is also used to locate Web services.
- WSDL is a W3C recommendation.

6.3.2 History

WSDL 1.0 (September 2000) has been developed by IBM, Microsoft and Ariba to describe Web Services for their SOAP toolkit. They built this by combining two service description languages: NASSL (Network Application Service Specification Language) from IBM and SDL (Service Description Language) from Microsoft.

WSDL 1.1, published in March 2001, is the formalization of WSDL 1.0. No major changes were introduced between 1.0 and 1.1.

WSDL 1.2 (June 2003) is still a working draft at W3C. According to W3C: WSDL 1.2 is easier and more flexible for developers than the previous version. WSDL 1.2 attempts to remove non-interoperable features and also defined the better HTTP 1.1 binding. WSDL 1.2 was not supported by most of the SOAP servers/vendors.

WSDL 2.0 became a W3C recommendation on June 2007. WSDL 1.2 was renamed to WSDL 2.0 because it has substantial differences from WSDL 1.1. The changes are:

- Adding further semantics to the description language.
- Removal of message constructs.
- No support for operator overloading.
- Port Types renamed to interfaces.
- Ports renamed to endpoints.

6.3.3 Description

The current version of the specification is 2.0; version 1.1 has not been endorsed by the W3C but version 2.0 is a W3C recommendation. WSDL 1.2 was renamed WSDL 2.0 because of its substantial differences from WSDL 1.1.

By accepting binding to all the HTTP request methods, WSDL 2.0 specification offers better support for Representational State Transfer (REST)

Web services, and is much simpler to implement. However support for this specification is still poor in software development kits for Web Services which often offer tools only for WSDL 1.1.

- WSDL defines services as collections of network endpoints, or ports.
- The WSDL specification provides an XML format for documents for this purpose.
- The abstract definitions of ports and messages are separated from their concrete use or instance, allowing the reuse of these definitions.
- A port is defined by associating a network address with a reusable binding, and a collection of ports defines a service.
- Messages are abstract descriptions of the data being exchanged, and port types are abstract collections of supported operations.
- The concrete protocol and data format specifications for a particular port type constitutes a reusable binding, where the operations and messages are then bound to a concrete network protocol and message format.

WSDL is often used in combination with SOAP and an XML Schema to provide Web services over the Internet. A client program connecting to a Web service can read the WSDL to determine what operations are available on the server. Any special data types used are embedded in the WSDL file in the form of XML Schema. The client can then use SOAP to actually call one of the operations listed in the WSDL.

6.3.4 Objects in a WSDL 1.1/WSDL 2.0

Service/Service: The service can be thought of as a container for a set of system functions that have been exposed to the Web-based protocols.

Port/Endpoint: The port does nothing more than defining the address or connection point to a Web service. It is typically represented by a simple http URL string.

Binding/Binding: Specifies the interface, defines the SOAP binding style (RPC/Document) and transport (SOAP Protocol). The binding section also defines the operations.

Port Type/Interface: The <portType> element, which has been renamed to <interface> in WSDL 2.0, defines a Web service, the operations that can be performed, and the messages that are used to perform the operation.

Operation/Operation: Each operation can be compared to a method or function call in a traditional programming language. Here the SOAP actions are defined and the way the message is encoded for example, 'literal'.

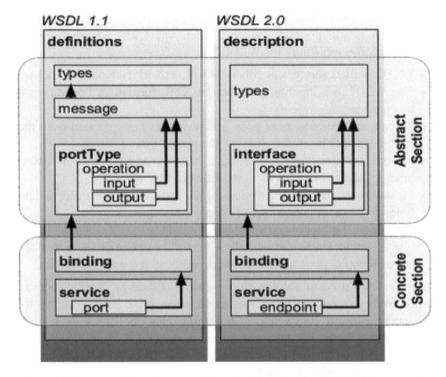

Figure 6.1 Representation of concepts defined by WSDL 1.1 and WSDL 2.0 documents.

Message/N.A.: Typically, a message corresponds to an operation. The message contains the information needed to perform the operation. Each message consists of one or more logical parts. Each part is associated with a message-typing attribute. The message name attribute provides a unique name among all messages. The part name attribute provides a unique name among all the parts of the enclosing message.

Parts are a description of the logical content of a message. In RPC binding, a binding may reference the name of a part in order to specify binding-specific information about the part. A part may represent a parameter in the message; the bindings define the actual meaning of the part. Messages had been removed in WSDL 2.0, where you simply and directly refer to XML schema types for defining bodies of inputs, outputs and faults.

Types/Types: The purpose of the types in WSDL is to describe the data. XML Schema is used (inline or referenced) for this purpose.

6.3.5 The WSDL Document Structure

A WSDL document describes a Web service using these major elements:

Element	Defines
`<types>`	The data types used by the Web service
`<message>`	The messages used by the Web service
`<portType>`	The operations performed by the Web service
`<binding>`	The communication protocols used by the Web service

The main structure of a WSDL document looks like this:

```
<definitions>

<types>
  definition of types........
</types>

<message>
  definition of a message....
</message>

<portType>
  definition of a port.......
</portType>

<binding>
  definition of a binding....
</binding>

</definitions>
```

A WSDL document can also contain other elements, like extension elements, and a service element that makes it possible to group together the definitions of several Web services in one single WSDL document.

6.3.5.1 WSDL Ports

The `<portType>` element is the most important WSDL element. It describes a Web service, the operations that can be performed, and the messages that are involved.

The `<portType>` element can be compared to a function library (or a module, or a class) in a traditional programming language.

6.3.5.2 Operation Types

The request-response type is the most common operation type, but WSDL defines four types:

Type	Definition
One-way	The operation can receive a message but will not return a response
Request-response	The operation can receive a request and will return a response
Solicit-response	The operation can send a request and will wait for a response
Notification	The operation can send a message but will not wait for a response

6.3.5.3 WSDL Messages

The `<message>` element defines the data elements of an operation. Each message can consist of one or more parts. The parts can be compared to the parameters of a function call in a traditional programming language.

6.3.5.4 WSDL Types

The `<types>` element defines the data types that are used by the Web service. For maximum platform neutrality, WSDL uses XML Schema syntax to define data types.

6.3.5.5 WSDL Bindings

The `<binding>` element defines the message format and protocol details for each port.

The binding element has two attributes: name and type. The name attribute (you can use any name you want) defines the name of the binding, and the type attribute points to the port for the binding, in this case the 'glossaryTerms' port.

The soap: the binding element has two attributes: style and transport.

The operation: element defines each operation that the port exposes.

6.3.6 WSDL Example

This is a simplified fraction of a WSDL document:

```
<message name="getTermRequest">
  <part name="term" type="xs:string"/>
```

```
</message>

<message name="getTermResponse">
  <part name="value" type="xs:string"/>
</message>

<portType name="glossaryTerms">
  <operation name="getTerm">
    <input message="getTermRequest"/>
    <output message="getTermResponse"/>
  </operation>
</portType>
```

In this example the `<portType>` element defines 'glossaryTerms' as the name of a port, and 'getTerm' as the name of an operation. The 'getTerm' operation has an input message called 'getTermRequest' and an output message called 'getTermResponse'. The `<message>` elements define the parts of each message and the associated data types. Compared to traditional programming, glossaryTerms is a function library, 'getTerm' is a function with 'getTermRequest' as the input parameter, and getTermResponse as the return parameter.

6.4 Universal Description Discovery and Integration (UDDI)

Universal Description, Discovery, and Integration (UDDI, pronounced Yu-di) is an XML-based registry for businesses worldwide to list themselves on the Internet. Its ultimate goal is to streamline online transactions by enabling companies to find one another on the Web and make their systems interoperable for e-commerce. UDDI is often compared to a telephone book's white, yellow, and green pages. The project allows businesses to list themselves by name, product, location, or the Web services they offer.

6.4.1 What Is UDDI?

UDDI is a platform-independent, XML-based registry for businesses worldwide to list themselves on the Internet. UDDI is an open industry initiative, sponsored by the Organization for the Advancement of Structured Information Standards (OASIS), enabling businesses to publish service listings and discover each other and define how the services or software applications interact over the Internet.

UDDI was originally proposed as a core Web service standard. It is designed to be interrogated by SOAP messages and to provide access to Web Services Description Language (WSDL) documents describing the protocol bindings and message formats required to interact with the Web services listed in its directory.

UDDI is a platform-independent framework for describing services, discovering businesses, and integrating business services by using the Internet.

- UDDI stands for Universal Description, Discovery and Integration.
- UDDI is a directory for storing information about Web services.
- UDDI is a directory of Web service interfaces described by WSDL.
- UDDI communicates via SOAP.
- UDDI is built into the Microsoft .NET platform [3].

6.4.2 History of UDDI

Microsoft, IBM, and Ariba spearheaded UDDI. The project now includes 130 companies, including some of the biggest names in the corporate world. Compaq, American Express, SAP AG, and Ford Motor Company are all committed to UDDI, as is Hewlett-Packard, whose own XML-based directory approach, called e-speak, is now being integrated with UDDI.

While the group does not refer to itself as a standards body, it does offer a framework for Web services integration. The UDDI specification utilizes World Wide Web Consortium (W3C) and Internet Engineering Task Force (IETF) standards such as XML, HTTP, and Domain Name System (DNS) protocols. It has also adopted early versions of the proposed Simple Object Access Protocol (SOAP) messaging guidelines for cross platform programming.

In November 2000, UDDI entered its public beta-testing phase. Each of its three founders – Microsoft, IBM, and Ariba – now operates a registry server that is interoperable with servers from other members. As information goes into a registry server, it is shared by servers in the other businesses. The UDDI beta is scheduled to end in the first quarter of 2001. In the future, other companies will act as operators of the UDDI Business Registry. UDDI registration is open to companies worldwide, regardless of their size.

6.4.3 Benefits of UDDI

Any industry or businesses of all sizes can benefit from UDDI. Before UDDI, there was no Internet standard for businesses to reach their customers and partners with information about their products and services. Nor was there a method of how to integrate into each other's systems and processes.

Problems the UDDI specification can help to solve:

- Making it possible to discover the right business from the millions currently online.
- Defining how to enable commerce once the preferred business is discovered.
- Reaching new customers and increasing access to current customers.
- Expanding offerings and extending market reach.
- Solving customer-driven need to remove barriers to allow for rapid participation in the global Internet economy.
- Describing services and business processes programmatically in a single, open, and secure environment.

6.4.4 UDDI Business Registration Components

- *White Pages* – address, contact, and known identifiers.
- *Yellow Pages* – industrial categorizations based on standard taxonomies.
- *Green Pages* – technical information about services exposed by the business.

6.4.4.1 White Pages

White pages give information about the business supplying the service. This includes the name of the business and a description of the business – potentially in multiple languages. Using this information, it is possible to find a service about which some information is already known (for example, locating a service based on the provider's name).

Contact information for the business is also provided – for example the businesses address and phone number; and other information such as the Dun & Bradstreet Universal Numbering System number. Thus this contains the basic contact information for each Web service listing. It generally includes basic information about the company, as well as how to make contact.

6.4.4.2 Yellow Pages

Yellow pages provide a classification of the service or business, based on standard taxonomies. These include the Standard Industrial Classification (SIC), the North American Industry Classification System (NAICS), or the United Nations Standard Products and Services Code (UNSPCS).

This has more details about the company, and includes descriptions of the kind of electronic capabilities the company can offer to anyone who wants to do business with it. It uses commonly accepted industrial categorization schemes, industry codes, product codes, business identification codes and the like to make it easier for companies to search through the listings and find exactly what they want. A single business may provide a number of services, there may be several Yellow Pages associated with one White Page.

6.4.4.3 Green Pages

Green pages are used to describe how to access a Web Service, with information on the service bindings. Some of the information is related to the Web Service – such as the address of the service and the parameters, and references to specifications of interfaces. Other information is not related directly to the Web Service – this includes e-mail, FTP, CORBA and telephone details for the service. Because a Web Service may have multiple bindings, a service may have multiple Green Pages, as each binding will need to be accessed differently.

6.4.4.4 UDDI Data

This specification presents an information model composed of instances of persistent data structures called entities. Entities are expressed in XML and are persistently stored by UDDI nodes. Each entity has the type of its outermost XML element. A UDDI information model is composed of instances of the following entity types:

- Business Entity: Describes a business or other organization that typically provides Web services.
- Business Service: Describes a collection of related Web services offered by an organization described by a business Entity.
- Binding Template: Describes the technical information necessary to use a particular Web service.
- TModel: Describes a 'technical model' representing a reusable concept, such as a Web service type, a protocol used by Web services, or a category system.
- Publisher Assertion: Describes, in the view of one business Entity, the relationship that the business Entity has with another business Entity.
- Subscription: Describes a standing request to keep track of changes to the entities described by the subscription.

6.4.5 UDDI Services and API Sets

This specification presents APIs that standardize behavior and communication with and between implementations of UDDI for the purposes of manipulating UDDI data stored within those implementations. The APIs are grouped into the following API sets.

6.4.5.1 Node API Sets

- UDDI Inquiry.
- UDDI Publication.
- UDDI Security.
- UDDI Custody Transfer.
- UDDI Subscription.
- UDDI Replication.

6.4.5.2 Client API Sets

- UDDI Subscription Listener.
- UDDI Value Set.

6.4.6 UDDI Nodes

A set of Web services supporting at least one of the Node API sets is referred to as a UDDI node. A UDDI node has these defining characteristics:

1. A UDDI node supports interaction with UDDI data through one or more UDDI API sets.
2. A UDDI node is a member of exactly one UDDI registry.
3. A UDDI node conceptually has access to and manipulates a complete logical copy of the UDDI data managed by the registry of which it is a part. Moreover, it is this data which is manipulated by any query and publish APIs supported by the node. Typically, UDDI replication occurs between UDDI nodes which reside on different systems in order to manifest this logical copy in the node.

6.4.7 UDDI Registries

One or more UDDI nodes may be combined to form a UDDI Registry. The nodes in a UDDI registry collectively manage a particular set of UDDI data. This data is distinguished by the visible behavior associated with the entities contained in it.

A UDDI Registry has the following defining characteristics:

1. A registry is comprised of one or more UDDI nodes.
2. The nodes of a registry collectively manage a well-defined set of UDDI data. Typically, this is supported by the use of UDDI replication between the nodes in the registry which reside on different systems.
3. A registry MUST make a policy decision for each policy decision point. It MAY choose to delegate policy decisions to nodes.

6.4.7.1 Affiliations of Registries

A UDDI registry affiliation has the following defining characteristics:

1. The registries share a common namespace for entity keys.
2. The registries have compatible policies for assigning keys to entities.
3. The policies of the registries permit publishers to assign keys.

6.4.7.2 Person, Publisher and Owner

When publishing information in a UDDI registry the information becomes part of the published content of the registry. During publication of an item of UDDI information, a relationship is established between the publisher, the item published and the node at which the publish operation takes place.

6.4.8 UDDI Representation

6.4.8.1 Representing Information within UDDI

For Web services to be meaningful there is a need to provide information about them beyond the technical specifications of the service itself. Central to UDDI's purpose is the representation of data and metadata about Web services. A UDDI registry, either for use in the public domain or behind the firewall, offers a standard mechanism to classify, catalog and manage Web services, so that they can be discovered and consumed. Whether for the purpose of electronic commerce or alternate purposes, businesses and providers can use UDDI to represent information about Web services in a standard way such that queries can then be issued to a UDDI Registry – at design-time or run-time – that address the following scenarios:

- Find Web services implementations that are based on a common abstract interface definition.
- Find Web service providers that are classified according to a known classification scheme or identifier system.
- Determine the security and transport protocols supported by a given Web service.

- Issue a search for services based on a general keyword.
- Cache the technical information about a Web service and then update that information at run-time.

6.4.8.2 Representing Businesses and Providers with 'Business Entity'

One top-level data structure within UDDI is the business Entity structure, used to represent businesses and providers within UDDI. It contains descriptive information about the business or provider and about the services it offers. This would include information such as names and descriptions in multiple languages, contact information and classification information. Service descriptions and technical information are expressed within a business entity by contained business service and binding template structures.

While the name of XML entity itself has the word business embedded in it, the structure can be used to model more than simply a 'business' in its common usage. As the top-level entity, business Entity can be used to model any 'parent' service provider, such as a department, an application or even a server. Depending on the context of the data in the entire registry, the appropriate modeling decisions to represent different service providers can vary [3].

6.4.8.3 Representing Services with 'Business Service'

Each business Service structure represents a logical grouping of Web services. At the service level, there is still no technical information provided about those services rather, this structure allows the ability to assemble a set of services under a common rubric. Each business service is the logical child of a single business entity. Each business service contains descriptive information – again, names, descriptions and classification information – outlining the purpose of the individual Web services found within it. For example, a business service structure could contain a set of Purchase Order Web services (submission, confirmation and notification) that are provided by a business.

Similar to the business Entity structure, the term business is embedded within the name business service. However, a suite of services need not be tied to a business, but can rather be associated with a provider of services, given a modeling scenario that is not based on a business use case.

6.4.8.4 Representing Web Services with 'Binding Template'

Each binding Template structure represents an individual Web service. In contrast with the business service and business entity structures, which are oriented toward auxiliary information about providers and services, a binding template provides the technical information needed by applications to bind and interact with the Web service being described. It must contain either the access point for a given service or an indirection mechanism that will lead one to the access point.

Each binding template is the child of a single business service. The containing parents, a binding template can be decorated with metadata that enable the discovery of that binding template, given a set of parameters and criteria.

6.4.9 Technical Models (tmodels)

Technical Models, or tmodels for short, are used in UDDI to represent unique concepts or constructs. They provide a structure that allows re-use and, thus, standardization within a software framework.

The use of tModels is essential to how UDDI represents data and metadata. The UDDI specification defines a set of common tModels that can be used canonically to model information within UDDI. If a concept that is required to model a particular scenario does not exist in a registry, a user should introduce that concept by saving a tModel containing the URL of the relevant overview documents: http://www.uddi.org/pubs/uddi_v3.htm#_Toc85907970.

6.5 Electronic Business XML Initiative (ebXML)

EbXML (Electronic Business using eXtensible Markup Language), is a modular suite of specifications that enables enterprises of any size and in any geographical location to conduct business over the Internet. Using ebXML, companies now have a standard method to exchange business messages, conduct trading relationships, communicate data in common terms and define and register business processes.

6.5.1 What Is ebXML?

EbXML was started in 1999 as an initiative of OASIS and the United Nations/ECE agency CEFACT. The original project envisioned and delivered five layers of substantive data specification, including XML standards for:

- Business processes.
- Core data components.

- Collaboration protocol agreements.
- Messaging.
- Registries and repositories.

6.5.1.1 ebXML Value

- Provides the only globally developed open XML-based Standard built on a rich heritage of electronic business experience.
- Creates a Single Global Electronic Market Enables all parties irrespective of size to engage in Internet-based electronic business. Provides for plug and play shrink-wrapped solutions.
- Enables parties to complement and extend current EC/EDI investment expand electronic business to new and existing trading partners.
- Facilitates convergence of current and emerging XML efforts.

6.5.1.2 ebXML Value Delivery

EbXML delivers the Value by

- Developing technical specifications for the open ebXML infrastructure.
- Creating the technical specifications with the world's best experts.
- Collaborating with other initiatives and standards development organizations.
- Building on the experience and strengths of existing EDI knowledge.
- Enlisting industry leaders to participate and adopt ebXML infrastructure.
- Realizing the commitment by ebXML participants to implement the ebXML technical specifications.

6.5.2 History of EbXML

The vision of ebXML is to create a single global electronic marketplace where enterprises of any size and in any geographical location can meet and conduct business with each other through the exchange of XML-based messages. EbXML enables anyone, anywhere, to do business with anyone else over the internet.

EbXML is a set of specifications that together enable a modular, yet complete electronic business framework. If the Internet is the information highway for electronic business, then ebXML can be thought of as providing the on ramps, off ramps, and the rules of the road.

EbXML is global since it has support, scope and implementation. It is a joint initiative of the United Nations (UN/CEFACT) and OASIS, developed with global participation for global usage. Membership in ebXML is open to anyone and the initiative enjoys broad industry support with over 75 member companies, and in excess of 2,000 participants drawn from over 30 countries. Participants represent major vendors and users in the IT industry and leading vertical and horizontal industry associations.

EbXML is evolutionary, not revolutionary. It is based on Internet technologies using proven, public standards such as: HTTP, TCP/IP, mime, SMTP, ftp, UML, and XML. The use of public standards yields a simple and inexpensive solution that is open and vendor-neutral. EbXML can be implemented and deployed on just about any computing platform and programming language.

Electronic commerce is not a new concept. For the past 25 years, companies have been exchanging information with each other electronically, based on Electronic Data Interchange (EDI) standards. Unfortunately, EDI currently requires significant technical expertise, and deploys tightly coupled, inflexible architectures. While it is possible to deploy EDI applications on public networks, they are most often deployed on expensive dedicated networks to conduct business with each other. As a result, EDI adoption has been limited to primarily large enterprises and selected trading partners, which represents a small fraction of the world's business entities. By leveraging the efforts of technical and business experts, and applying today's best practices, ebXML aims to remove these obstacles. This opens the possibility of low cost electronic business to virtually anyone with a computer and an application that is capable of reading and writing ebXML messages. Businesses of all sizes will adopt ebXML for reasons of lower development cost, flexibility, and ease of use.

The Messaging Service, Registry & Repository, and Collaborative Partner Agreement, comprise the core ebXML infrastructure specifications. The Messaging Service specification has gone through several drafts and proof of concept demonstrations and is stable enough to be used for early development work.

EbXML is an end-to-end solution, but it is modular in nature. Businesses are not required to implement the entire specification all at once. Specification sets may be implemented incrementally as needed. The specification sets, each being stand-alone, but all of them being aligned with each other, can be implemented individually or in a number of useful combinations.

As the world adopts ebXML, barriers to participation in global electronic business will fall away. Virtually free transport will be available over the Internet. Standards-based business process and message definitions will drastically reduce the cost of negotiating trading partner agreements. EbXML compliant off-the-shelf software will enable any company, however low-tech, to easily implement their participation in ebXML-based electronic business.

The ebXML enabled world is one of mixed, small, medium- and large-sized enterprises conducting business with each other in much more standard and low cost ways than today. Numerous opportunities will exist for all businesses to move to new, innovative, service-based business models, without having to fundamentally change existing architectures or software.

6.5.3 Architectural Overview of ebXML

The ebXML architecture provides:

1. A way to define business processes and their associated messages and content.
2. A way to register and discover business process sequences with related message exchanges.
3. A way to define company profiles.
4. A way to define trading partner agreements.
5. A uniform message transport layer.

The ebXML initiative is designed for electronic interoperability, allowing businesses to find each other, agree to become trading partners and conduct business. All of these operations can be performed automatically, minimizing, and in most cases completely eliminating the need for human intervention. This streamlines electronic business through a low cost, open, standard mechanism.

In order for enterprises to conduct electronic business with each other they must:

- Discover each other and the products and services they have to offer.
- Determine which shared business processes, and associated document exchanges, to use for obtaining products or services from each other.
- Determine the contact points and form of communication for the exchange of information.
- Agree on the contractual terms on the above chosen processes and associated information.
- Exchange information and services in an automated fashion in accordance with these agreements.

ebXML is designed to meet these needs and is built on three basic concepts: provide an infrastructure that ensures data communication interoperability;

provide a semantics framework that ensures commercial interoperability; and provide a mechanism that allows enterprises to find each other, agree to become trading partners and conduct business with each other.

1. Infrastructure to ensure data communication interoperability is provided through:

   (a) a standard message transport mechanism with a well defined interface, packaging rules, and a predictable delivery and security model;

   (b) a 'business service interface' that handles incoming and outgoing messages at either end of the transport.

2. Semantic Framework to ensure commercial interoperability is provided through:

   (a) metamodel for defining business process and information models;

   (b) set of re-useable business logic based on core components that reflect common business processes and XML vocabularies;

   (c) Process for defining actual message structures and definitions as they relate to the activities in the Business Process model.

3. Mechanism to allow enterprises to find each other, agree to establish business relationships, and conduct business, is provided through:

   (a) shared repository where enterprises can register and discover each other's business services via partner profile information;

   (b) process for defining and agreeing to a formal Collaboration Protocol Agreement (CPA), if so desired or where required;

   (c) shared repository for company profiles, business process models and related message structures.

   The technical architecture is composed of five main area of emphasis:

1. Business Process and Information Model.
2. Company Profiles.
3. Messaging Services.
4. Registry & Repository.
5. Collaborative Partner Agreements.

The Business Process models define how business processes are described. Business Processes represent the 'verbs' of electronic business and can be represented using modeling tools. The specification for business process definition enables an organization to express its business processes so that

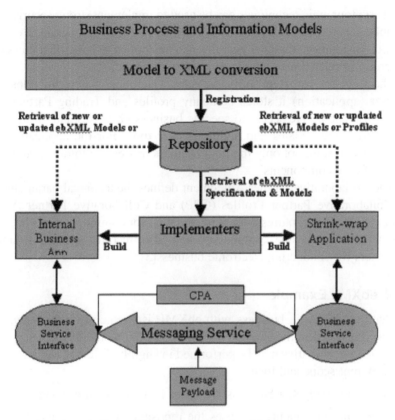

Figure 6.2 ebXML technical architecture.

they are understandable by other organizations. This enables the integration of business processes within a company, or between companies.

The Information models define reusable components that can be applied in a standard way within a business context. These Core Components represent the 'nouns and adjectives' of electronic business. They are defined using identity items that are common across all businesses. This enables users to define data that is meaningful to their business while also maintaining interoperability with other business applications.

The ebXML Messaging Service specification defines the set of services and protocols that enables electronic business applications to exchange data. The specification allows any application-level protocol to be used. These can include common protocols such as SMTP, HTTP, and FTP. Well established cryptographic techniques can be used to implement strong security. For example, secure protocols such as HTTPS can be used to guarantee confidentiality. In addition, digital signatures can be applied to individual messages or a group of related messages to guarantee authenticity.

The Registry and Repository provides a number of key functions. For the user (application) it stores company profiles and Trading Partner specifications. These give access to specific business processes and information models to allow updates and additions over time. For the application developer it will store not only the final business process definitions, but also a library of core components.

The Collaborative Partner Agreement defines the technical parameters of the Collaborative Partner Profiles (CPP) and Collaborative Partner Agreements (CPA). This captures critical information for communications between applications and business processes and also records specific technical parameters for conducting electronic business[3].

6.5.4 ebXML Example

Conducting electronic business with ebXML is essentially a four step process, as illustrated in the example below. There are many different scenarios in which these activities may be performed in slightly different sequences and with different scope and focus.

1. Design and register business processes and information models

   - The implementer browses the repository for appropriate business processes, or for the process the intended partner is registered to support.

2. Implement business service interfaces and register Collaborative Partner Profiles

   - The implementer buys, builds, or configures application(s) capable of participating in the selected business process.
   - The implementer registers his (software's) capability to participate, in the form of a Collaborative Partner Profile.

3. Optionally negotiate and define a Collaborative Partner Agreement (CPA)

   - The two parties negotiate technical details and/or functional overrides, and draw up the result in the form of a CPA.
   - Parties optionally register the CPA.

4. Exchanging messages between business partners.

- The parties (software) send and receive ebXML messages containing ebXML business documents, over the secure and reliable ebXML Messaging Service.

6.6 Summary

The vision of ebXML is to create a single global electronic marketplace where enterprises of any size and in any geographical location can meet and conduct business with each other through the exchange of XML-based messages. EbXML enables anyone, anywhere, to do business with anyone else over the internet.

EbXML is a complete set of specifications to enable secure, global, electronic business using proven, open standards such as TCP/IP, HTTP, and XML. ebXML is also evolutionary in nature, built on 25 years of EDI experience, designed to work with existing EDI solutions, or be used to develop an emerging class of Internet-based electronic business applications based on XML.

Since system integration and software interoperability are the cornerstones of any successful IT infrastructure, ebXML is built on an infrastructure that ensures electronic interoperability. This is accomplished by providing an open semantics framework that allows enterprises to find each other, agree to become trading partners, and conduct business. The evolution of many new business models will be enabled by ebXML, through business process patterns and the 'commoditization' of such business processes.

The electronic business infrastructure provided by ebXML is broad in scope and well integrated, providing the best alternative in the industry today. And perhaps most importantly, ebXML is platform and vendor neutral, providing an industry solution based on open standards, designed through a collaborative and open process.

References

1. Chinese School. An Introduction to SOAP (Simple Object Access Protocol). http://chinese-school.netfirms.com/computer-article-SOAP.html, 2009.
2. Jennifer Kyrin. Simple Object Access Protocol – SOAP. Beethany, http://webdesign.about.com/od/soap/a/what-is-xml-soap.htm, 2010.
3. OASIS Open. Enabling Electronic Business with ebXML. ebXML, http://www.ebxml.org/white_papers/whitepaper.htm, 2000.

4.   OASIS Open. UDDI Spec TC. http://www.uddi.org/pubs/uddi_v3.htm#_Toc85907970, 2004.
5.   Techtarget. SOAP. http://searchsoa.techtarget.com/sDefinition/0,,sid26_gci214295,00.html, 2010

7

Integrating SOA and Web Services Introduction

7.1 Introduction

Service-Oriented Architectures (SOAs) and Web Services combine to provide an opportunity for organizations to reduce the costs and complexities of application integration inside the firewall and open up new possibilities for legacy applications to participate in eBusiness. Integration is reuse of data and functionality across applications, services and enterprises. The enterprise uses different applications for running its business. Lack of integration among applications leads to scattered data islands and duplicate data. Integration tries to unlock value by integrating applications, thus helping organizations to leverage their legacy applications and maximize the ROI for existing investments (integration). Integration based on services creates the stepping-stones for the SOA journey. Integration based on SOA principles and infrastructure goes a long way toward reaching business goals and building an IT eco system.

7.2 Overview of Integration

SOA is essentially a distributed architecture, with systems that span computing platforms, data sources, and technologies. A distributed architecture requires integration. By standardizing how systems interoperate, Web services simplify the task of integration. Web services alone, however, do not suffice. Organizations need an evolutionary approach to SOA that incorporates legacy (non-Web-services-based) systems. Integration software provides the bridge between the legacy systems and SOA, allowing organizations to leverage existing software assets while managing their transition to SOA. Integration solutions also contribute mature technologies – such as messaging, routing, data translation and transformation, and event management – along with organizational disciplines that are necessary for full-fledged, enterprise SOA. Moreover, integration capabilities such as business process management (BPM) and business activity monitoring (BAM) allow organizations to realize a higher level of business productivity from SOA by enabling the optimization of business processes and the alignment of strategic objectives with operational actions. In essence, integration should play a central role in any organization's SOA strategy.

7.2.1 Need for Integration

Business cycles, operations, and responsiveness are constrained by several parameters in the enterprise. A few important and key root causes are

- Unmanaged, non-upgradeable, non-scalable applications.
- Non-integrated stovepipe applications.
- Complex, obscure, and inefficient business processes.
- Mergers, acquisitions, and regulations.
- Deviations from the enterprise strategic objectives in daily operations.

Integration technology can play a significant role in enabling the parameters described above by providing the vital glue whereby enterprises can overcome bottlenecks and make the businesses more responsive, eventually facilitating an early go-to-market. As a result, Integration technology will facilitate leveraging the existing assets while creating new business opportunities.

7.2.2 Integration Technologies

Different integration technologies and infrastructure services are needed to accomplish different integration use cases. Therefore, enterprise will inevitably acquire a collection of technologies to meet their diverse business needs.

7.2.2.1 Application Integration Services

Application integration services offer the basic features and capabilities required to integrate and connect applications reliably and securely. These services include

- Connectivity to the applications. Connectivity is generally achieved using adapters, which enable a standard application connectivity approach and eliminate the need to program against proprietary application

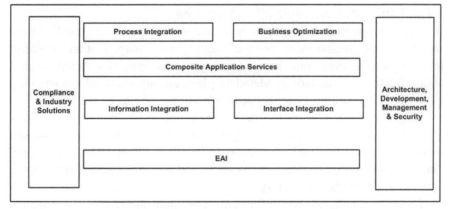

Figure 7.1 Integration technologies.

program interfaces (APIs). If Web services interfaces are available, application connectivity may be achieved without an adapter. Application connectivity may also be achieved at the database level using a database adapter or a direct SQL interface.

- Data transport and routing capabilities to move information reliably and securely between systems. Application integration solutions typically use a message-based approach to moving data, and they support a variety of topologies that link systems together in a scalable and flexible manner.
- Data transformation and mapping capabilities. These enable data to be converted to and from the required target representations.
- Infrastructure services to implement the application integration scenarios include support for transactions, application-level security, and exception management services.

The core application integration capabilities often play a role in other integration services, such as information integration, interface integration, composite applications, process integration, and business optimization with business activity monitoring (BAM). Without the underlying application integration services, these other integration services require point-to-point connectivity with backend applications, which limits agility and reuse. Technologies that provide application integration services include

- Message brokers.
- Integration servers.
- Enterprise service buses (ESBs).

7.2.2.2 Information Integration Services

Information integration services consolidate and integrate structured and unstructured information from multiple sources, and manage it at an enterprise level to promote data consistency. The key to managing distributed information effectively is metadata. Metadata plays an important role in enabling interoperability among systems because it allows data to be understood and exchanged consistently across the systems. Enterprise metadata repositories that provide real-time data services enable reuse. The next step in data management services and integration productivity is semantic integration. Semantic integration captures the meaning and context of data within source systems in a metadata repository.

Data integration involves a framework of applications, tools, techniques, technologies, and management services for providing a unified and consistent view of enterprise business data to business processes and business users. Applications are custom-built or vendor-developed solutions that utilize one or more data integration tools.

- Tools are off-the-shelf commercial products that support one or more data integration technologies. These tools are used to design and build data integration applications.
- Technologies implement one or more data integration techniques.
- Techniques are technology-independent approaches for doing data integration.
- Management services support the management of data quality, metadata, and data integration system operations.

7.2.2.3 Data Integration Techniques

The three main techniques used for integrating data are

1. Data consolidation.
2. Data federation.
3. Data propagation.

These three techniques may, in turn, use changed data capture and data transformation techniques during data integration processing.

Data Consolidation

Data consolidation captures data from multiple source systems and integrates it into a single persistent data store. Essentially, with data consolidation, there will be a delay, or latency, between the time updates occur in source systems and the time those updates appear in the target store. Depending on business needs, this latency may be a few seconds, several hours, or many days.

Data Federation

Data federation provides a single virtual view of one or more source data files. When a business application issues a query against this virtual view, a data federation engine retrieves data from the appropriate source data stores, integrates it to match the virtual view and query definition, and sends the results to the requesting business application. By definition, data federation always pulls data from source systems on an on-demand basis. Any required data transformation is done as the data is retrieved from the source data files.

Data Propagation

Data propagation applications copy data from one location to another. These applications usually operate online and push data to the target location; i.e., they are event-driven. Updates to a source system may be propagated asynchronously or synchronously to the target system.

7.2.2.4 Enterprise Content Management

Another technology that handles the integration of unstructured data is Enterprise Content Management (ECM), which is focused on the consolidation of documents, Web information, and rich media. ECM products concentrate on the sharing and management of large quantities of unstructured data for a wide user population. These products add a content management layer on top of a shared data store. This layer provides metadata management, versioning, templates, and workflow.

An ECM content store can act as a data source for an EII or ETL application. The key here is not simply to provide access to unstructured data, but also to access the metadata that describes the structure, contents, and business meaning of that data.

7.2.2.5 Interface Integration Services

Interface integration services provide access from a variety of front-end devices to a variety of back-end systems. The technologies that provide interface integration services include portals and mobile integration gateways.

Portals

Portals provide an integrated user interface to multiple back-end systems. Many EAI solutions focus on automating transactions across systems. But a portal solution integrates information into a browser for a particular purpose or set of users, providing a single touch point for the delivery of application services.

Mobile Integration Gateways

Mobile devices have different screen sizes, resolutions, and interface requirements. Mobile integration solutions usually include a specialized server that provides the transformation services for each target device, along with security. This server integrates with back-end systems either directly (point-to-point) or through an integration platform. The mobile integration server typically uses message queues and logs to guarantee that messages will be delivered once, and only once, even in the event of a system failure. To support occasional users, it controls message delivery by organizing messages into transactional groups, and determines when messages are delivered, based on the type of network connection. To address the insecurity of mobile communications, the ability to encrypt data is important [1].

7.2.2.6 Service Registries for Governance

UDDI is the registry standard for dynamically discovering and invoking Web Services. UDDI was originally created to manage the cataloging of public Web services. Registry products typically incorporate some level of Web services management capability to make the registry more useful within the enterprise.

7.2.2.7 Composite Application Services

The development, deployment, and management of composite applications would involve the following major list of activities:

- Develop new services.
- Create service abstractions by combining several low-level services into a more useful, reusable high-level business service.
- Design and model the behavior of the composite application, including how information and messages flow across services, and how services need to be orchestrated in support of a business process.
- Discover services or browse from a registry of available services.
- Manage policies and service level agreements to engender trust and enable the reuse of services.
- Deploy the resulting solution into a managed run-time environment.

- Technologies most commonly associated with composite applications include Web services, service orchestration capabilities, and service registries with effective governance.

7.2.2.8 Web Services

Web services define a standardized interface (Web Services Description Language, or WSDL), a standardized communication protocol (Simple Object Access Protocol, or SOAP), a standardized repository for registering and discovering Web services (Universal Description, Discovery, and Integration, known as UDDI), and standardized message encoding using XML. These standards enable a Web service to reside anywhere and be accessed from everywhere, making Web services well suited to the role of providing aggregated functionality within a composite application and being the standard for inter application interfaces.

7.2.2.9 Orchestration

To create a composite application from a set of Web services, it is necessary to define the flow of control across services and also the dialog control logic. This is achieved using service orchestration. This latter approach basically allows developers to modify the way services are linked together without changing the services themselves. Business Process Execution Language (BPEL), a vendor-led initiative, is gaining mind share and market share as the orchestration standard.

7.2.2.10 Process Integration Services

A process-driven approach to integration improves the alignment between IT and the organization. It starts with the development of business process models that business people can understand and review. These models depict a shared understanding of the end-to-end business process. Using process integration technology, the models can then be automated. This significantly reduces the chance of misinterpreting the business requirement or getting the process wrong. Since processes can span departments, business units, and organizations, no single individual may own or even understand the end-to-end process. An enterprise business processes may be automated, manual, or collaborative. Frequently, an end-to-end business process includes both automated and manual processes (workflow). From an integration perspective, each type of process has different requirements and therefore calls for different process technologies. The technologies that deliver process-level integration include business process management for handling automated (and sometimes manual) processes, workflow management for manual processes, and groupware or collaboration platforms.

7.2.2.11 Business Optimization Services

Business Optimization services enable the organizations to align business strategies with coordinated actions. The goal of the BPM is to streamline the processes, but optimizing individual processes does not necessarily lead to optimization at an enterprise level. The goal of the Business Optimization is to optimize at the enterprise level and to align all parts of the enterprise to enhance business performance. Business Optimization Services include the technologies that enable business performance management. Business Performance includes methodologies like Six Sigma, Balanced scorecard, Lean, and technology infrastructure.

7.2.2.12 Technology Choices

Two key characteristics of a business process are its dynamism (frequency of change), on the one hand, and the complexity of coordination the particular process requires on the other. Different types of processes are best addressed using different types of technologies, such as EAI, application servers, or BPM/Workflow.

7.3 Design and Development of SOA for Integration

SOA is essentially a distributed architecture, with systems that span computing platforms, data sources, and technologies. A distributed architecture requires integration. By standardizing how systems interoperate, Web services simplify the task of integration. Web services alone, however, do not suffice [2]. Organizations need an evolutionary approach to SOA that incorporates legacy (non-Web-services-based) systems. Integration software provides the bridge between the legacy systems and SOA, allowing organizations to leverage existing software assets while managing their transition to SOA.

Integration solutions also contribute mature technologies – such as messaging, routing, data translation and transformation, and event management – along with organizational disciplines that are necessary for full-fledged, enterprise SOA. Moreover, integration capabilities such as business process management (BPM) and business activity monitoring (BAM) allow organizations to realize a higher level of business productivity from SOA by enabling the optimization of business processes and the alignment of strategic objectives with operational actions. In essence, integration should play a central role in any organization's SOA strategy.

7.3.1 The Integration Landscape

Service-Oriented Architecture describes a set of standards-based technologies and a design approach to create interoperable systems. SOA principles can be used to create services, aggregate services into composite applications, develop whole new applications, and to integrate existing applications. Whether building new applications or integrating existing ones, the goals are to create interoperable, sustainable, robust solutions; solutions that meet today's business needs while being flexible enough to meet future needs. As SOA principles become more broadly applied, the SOA landscape becomes more diverse and heterogeneous, both from a technology and a semantic perspective. Accommodating a diverse landscape requires an integration-based architecture pattern where integration techniques are used to achieve a semantic consistency while allowing for and accepting broad-based heterogeneity. An integration-oriented approach allows us to treat applications and services as a heterogeneous collection of 'black boxes', to ignore their inner workings, and concentrate on their public interfaces. The integration layer with its transport and transformation services provides interoperability. This approach provides developers a significant degree of independence to evolve their systems.

The integration landscape includes all of the applications, databases, and data resources, as well as composite applications and any relevant trading partners and customer systems that need to interoperate within the context of the enterprise's business processes [3]. Finally, the landscape includes the ESBs and other integration layer services specifically designed to facilitate service-to-service communication, message exchange, and reconciliation. In the integration pattern, the role of the common model is to simplify the development of the 'shared data services'.

A metadata landscape covers the same domain as the integration landscape but includes only the metadata exposed in service interfaces; metadata that describes the information flowing through the landscape. Without a common model in the metadata landscape we have different schemas representing each document in each service. Without the common model, each interface requires its own data integrity rules and its own mapping to each service it is integrated with.

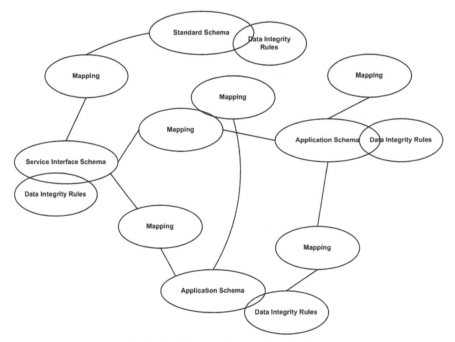

Figure 7.2 Point-to-point metadata landscape.

From a metadata landscape perspective, a common model dramatically reduces the number of mappings and overall complexity. A common model organizes all of integration metadata into a shared context with a shared set of semantics. With a common model, each interface is mapped once and the data integrity rules are defined only once.

7.3.2 Service Oriented Integration

Service Oriented Integration is a more application-agnostic approach for integration. It is a functional integration technique which leverages exposed business functionality as services. These services are basically published and accessed in a standard way. SOA promotes assembly, orchestration, and choreography based on service, made possible by a Service-based integration. Integration approaches used inside the enterprise have come a long way from the point-to-point customized integration approach to a more disparate Service Oriented Integration. An initial approach, with point-to-point integration can be termed more accurately an application-to application integration where each application owner may get involved in sorting out the

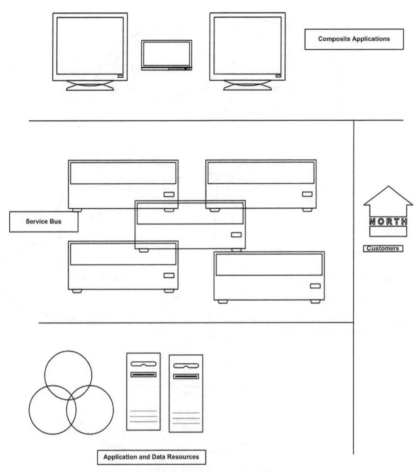

Figure 7.3 Metadata landscape with a common model.

difference in platform, data semantics, and communication protocols. Some design document is prepared for this integration and for extra lines of code based on the design. This process gives us, first, point-to-point integration. It requires a host for the application and the communication protocol and port number details need to be coded inside the program or read from outside config files. This approach becomes very cumbersome as the number of integration points increase. Basically uses EAI software and a myriad of adapters for legacy and ERP applications like SAP, Baan, Siebel, etc. It solves the point-to=point connection problem by providing central access to each integration application. It provides messaging, reliability, choice of transport

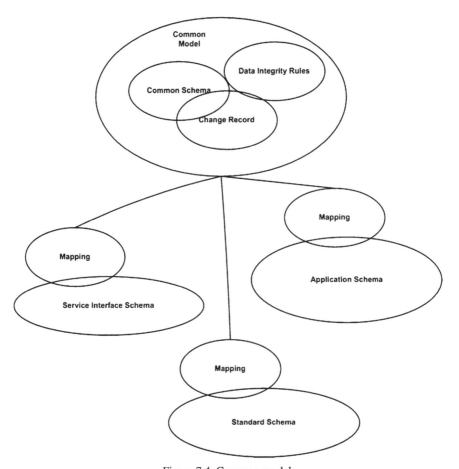

Figure 7.4 Common model.

and transformation. This approach is flawed by the need for skill resources, an inherent complexity, and the need for centralized control. The Service Oriented approach, on the other hand, proposes a standard way of integration based on services. It could be implemented via different technologies or patterns. In this approach, the designer is basically concerned with functionality and granularity, rather than plumbing issues between the applications. The underlying integration infrastructure should take care of interaction between the services. It provides a more uniform approach to all integration needs, like b2b, portal, data integration, and so on.

The diagram in Figure 7.5 depicts a SOA Integration.

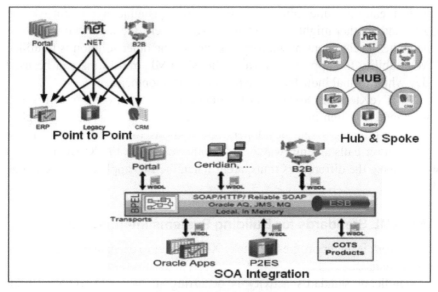

Figure 7.5 SOA integration.

7.4 The Role of XML and Web Services in SOA for Integration

XML is fast becoming the worldwide standard for exchanging data because it is human readable, self describing, flexible, platform-independent and widely supported. XML is a simple, very flexible text format derived from a language originally designed to meet the challenges of large-scale electronic publishing. It is written in plain text, usually English, making it easy to read, understand, and implement.

XML is also a data format. Like other common data formats (such as CSV, .dbf, .mdb, etc.), XML can be used in programs for calculations, sorts, queries, reports, displays, mergers of data, etc.

XML is not a programming language, nor is it a protocol. By itself, it does nothing. However, when data is stored and communicated in XML's text based format, programmers gain a powerful tool for the development of interoperability between applications.

7.4.1 Transformation of Data Using XML

One of the advantages of XML is that any software developer can define their own tags. However, this can also be a problem if two developers use different tags to mean the same thing. For instance one manufacturer might use the tag "" and another might use the tag " ". Fortunately, there are two solutions. One is the development of industry standards. The other solution is to utilize XML's ability to support 'Transformations'. XML transforms can be used today while formal industry standards are in development.

XSLT (eXtensible Style sheet Language Transformations), which is also conveniently written in XML, provides a way to 'map' or create a template that cross-references one developer's conventions to another's. So, if one manufacturer calls an entryway a and another calls it a, the XSLT transform would make the differences transparent when the two applications exchange data [4].

7.4.2 XML Standards for Building Systems Interoperability

Today, work is under progress to make XML easier to exchange their industry specific data.

Significant standards activity is occurring in the oBIX (Open Buildings Interface Exchange) technical committee of OASIS (Organization for the Advancement of Structured Information Standards). CABA (Continental Automated Buildings Association) served as an incubator for oBIX before it affiliated with OASIS for a closer association with the IT industry. The ASHRAE (American Society for Heating Refrigeration and Air conditioning Engineers) Standing Standard Project Committee 135 is actively developing an XML interface for their Data Communication Protocol for Building Automation and Control Networks (BACnet). SIA (the Security Industry Association) is also actively developing standards for interoperability.

While official standards for each industry will take time to develop, many vendors will start using the preliminary templates and constructs proposed from the committee work prior to that. The demand is high, the benefit is great, and changes are easy to make. It is eXtensible. At present, Microsoft is spending billions of dollars annually on XML and Web services because of the business opportunities these tools open up. Their Office 2003 package stores data natively in XML, and Internet Explorer is XML enabled.

7.4.3 The Interoperability Vision

Customers are now free to choose individual building control systems. As long as they choose systems that are truly open – like those based on XML and APIs with SDKs and professional services available – they will enjoy interfaces that are flexible, customer controlled, and yields a strong ROI.

In the not-so-distant future, end-users and systems integrators will be able to obtain off-the- shelf interoperability applications for specific manufacturers. Those applications will provide standard templates and mapping tools to join the silos of information that exist among the various systems within the enterprise. New solutions will emerge that result from interoperability among two or more applications.

7.4.4 Binary Protocols vs. XML

Today, most controllers communicate with devices or other controllers using a binary protocol. So, integration between two vendors means one has to acquire the binary protocol of the other and adapt to their proprietary, ever-changing rules. Industry standard protocols have overcome this problem to some extent.

Binary protocols (based on ones and zeros) have been around for decades. Although computers still talk in ones and zeroes, the gigabit Ethernet speeds available today make it practical to use human readable XML text for intranet and machine-to-machine (M2M) communications of data. After all, the Internet is built upon TCP/IP using HyperText Transport Protocol (HTTP), which is optimized for 'text'.

More importantly, binary protocols are not extensible. Once a binary protocol is established, it is not easily changed. Whether to describe the data on the magnetic stripe of your credit card or to define communications between controllers, a change would probably mean reissuing all credit cards, and updating or replacing controllers. With XML, another 'element' can be added as easily as another item can be added to an outline in a Word document.

7.5 Web Services Interoperability

Developing Web services has become challenging even for an experienced developer. Over the past few years, however, the Web services community has evolved and mostly agreed on some basic rules when developing Web services that, if followed, will produce the most interoperable Web services possible [5]. These best practices consist of three basic rules:

1. Use contract-first design principles.
2. Use the document/literal form of the Web Services Description Language.

3. Follow the WS-I Basic Profile 1.0.

When developing Web services, it is necessary to start with a WSDL (Web Services Description Language) file. Although WSDL files are simply XML, the WSDL 1.1 specification defines what valid XML can make up a service description. Much of a WSDL file has redundant code, which makes it tedious to write by hand. The five major components of the WSDL file are namespace declarations, types, messages, ports, and bindings. Of the major components of the service description the types usually take most of the work. It is here that the data types need to be defined so that Web services servers and clients can agree on what data is being used. Since the type's element consists of just an XML schema definition, however, XSD editors can be used to create the necessary element definitions. Microsoft's Visual Studio.NET has an XSD editor that allows the user to toggle between a graphical designer and a text editor that features IntelliSense. Altova's XMLSpy is another good editor for type elements and provides validation for all sorts of XML files, ensuring that no errors are accidentally introduced. Using such an editor is highly recommended because it allows the developer to create and modify XSD files quickly and easily [6].

The open source toolkits include no WSDL generators, each service description must be done by hand in a text editor. Here the commercial tools really shine, because once the schemas are defined for the service types, creating the WSDL file is as trivial as using a typical Windows wizard. Thinktecture's WsContractFirst tool for designing types, services, and methods, now known as WSCF, is an integral part of the Microsoft Visual Studio Environment. If the VS.NET Add-In is installed, right-clicking on XSD files in the solution explorer will allow the user to select 'Create WSDL Interface Description'. Clicking this will launch the wizard. The user will first be prompted for the service name, XML namespace, and optional documentation. The user can then specify all his services and indicate whether they are Request/Response or One-Way. The next step lets the user associate a type with the in-and-out methods of the operations just defined. The last step allows the user to specify another XSD location, if desired. The WSDL will then be generated automatically.

The WSDL file will consist of the four elements mentioned earlier: types, message, portType, and binding. However, one more element is needed to make it complete: service. If the user develops the service in VS.NET, the service element will automatically be added to the WSDL. However, it can just as easily be added by hand. It should look something like this: The other tactic to ensure that Web services are as interoperable as possible is to follow the Basic Profile 1.0. The Basic Profile 1.0 strongly suggests the use of the 'document/literal' form of service descriptions. It is a claim that the use of 'document/literal' is necessary, but not sufficient, to ensure interoperability.

7.6 J2EE and .NET Interoperability

Weaving together Web services to create cross-organizational business processes requires all partners to program to the same standard model and to avoid exposing proprietary implementations. After many years of promoting the interoperability among vendors through joint efforts on standardizing protocols, significant progress has been made. Web services technology offers the promise and hope of integrating disparate applications in a seamless fashion. But enterprise applications are built around different technologies and platforms, and integration across businesses is never a trivial task. The relatively recent emergence of Business Process Execution Language (BPEL) for Web services provides a higher-level description language to specify the behavior of Web services. It provides a standard and portable language for orchestrating Web services into business processes. As BPEL was embraced by major vendors, IDE tools designed to automate business process design, such as the IBM WebSphere Studio Application Developer Integration Edition (Application Developer), entered the marketplace.

The tools eliminate a large amount of required coding in the Web services integration, and allow user to construct business processes by dragging and dropping the WSDL files into the tools. It is expected that the tools automatically generate the client stubs for the participating Web services. As a result, the success of the integration is now largely dependent upon the underlying sophisticated tooling support. This puts an even greater demand on developers to adopt best practices that ensure the inherent interoperability of participating Web services. Appropriate attention should be provided to answer the following issues:

- Using vendor tools to derive the Web services semantics in WSDL from implementation code is convenient, but this approach ignores the design of the message schemas which is central to Web services interoperability in heterogeneous environments.
- The ease, flexibility, and familiarity of the popular RPC/encoded style makes it an attractive choice for developers, however, the difficulty in synchronizing the implementations of the abstract SOAP encoding data

model among vendors presents a difficult challenge for Web services interoperability.

- Weakly-typed collection objects, arrays containing null elements, and certain native data types all pose special problems for interoperability. Specifically:
    - It is impossible for vendor tools to accurately interpret XML Schemas representing weakly-typed collection objects and map them to the correct native data types.
    - The XML representations of an array with null elements differ between .NET and WebSphere.
    - Because native data types and XSD data types do not share a one-to-one mapping, information or precision can be lost during the translation.
    - Different naming conventions in .NET and Java technology can result in namespace conflicts, as can the use of relative URI references.

7.6.1 Interoperability and WSDL

All Web services interoperability issues center on the WSDL file. WSDL is the interface definition language (IDL) of Web services and the contract between the client and the server. The services semantics, namely the message types, data types, and interaction styles in the WSDL file, are the key to building loosely-coupled systems. Even though WSDL does not mandate the use of a particular type system, the XML Schema data type (XSD) is widely embraced by the Web services community. XSD offers a wide range of built-in primitive types and further allows service providers to define custom complex types. The XSD-type system is more sophisticated and powerful than the type system of any programming language, and more importantly it is language-neutral. That makes WSDL the logical starting point to define the Web services semantics.

Ideally, just like the IDLs for COM and CORBA, user should create and edit the WSDL first – define the interfaces, messages, and data types before you build the Web services and clients in specific implementation languages based on the service semantics in the WSDL. But programmers start with their implementation-language-specific interfaces, such as Java interfaces, and rely on vendors' tools to derive the service semantics in WSDL from their implementation code; they then leave it to the Web services client to figure out from the WSDL how to make the calls to the Web service. In the end, they often don't even need to know about the WSDL before they can build and run a sophisticated Web service. They might only have run their client and server in a homogeneous environment; that is, either on a J2EE platform or on a .NET platform, but not on both.

The danger becomes evident when there is a demand for interoperability across heterogeneous environments. Now to create the key artifacts for Web services interaction from implementation languages and then using the platform-dependent tools to map them to the language-neutral ones in WSDL [7]. When the tool on the client platform generates the service proxy from this tool-generated WSDL, another level of mapping occurs. This process deviates from the hard fact that WSDL is the IDL for Web services. The language-neutral WSDL should be the common ground for both client and server and using the tools in this way doubles the likelihood that information is lost during the mappings.

Automation has become an important aspect of today's enterprise application development landscape. The tools are powerful; it is a matter of how user's leverage their power. Tools can be used to generate a skeleton WSDL to serve as a starting point or as a template. Schemas, message parts, and data bindings must be carefully designed in accordance with the WS-I recommendations. WSDL is the single most important artifact for Web services interoperability.

7.6.2 Best Practices for Web Services Interoperability

- Design the XSD and WSDL first, and program against the schema and interface.
- If at all possible, avoid using the RPC/encoded style.
- Wrap any weakly-typed collection objects with simple arrays of concrete types as the signature for Web service methods.
- Avoid passing an array with null elements between Web services clients and servers.
- Do not expose unsigned numerical data types in Web services methods. Consider creating wrapper methods to expose and transmit the data types.
- Take care when mapping XSD types to a value type in one language and to a reference type in another. Define a complex type to wrap the value type and set the complex type to be null to indicate a null value.
- Because base URIs are not well-defined in WSDL documents, avoid using relative URI references in namespace declarations.

- To avoid conflicts resulting from different naming conventions among vendors, qualify each Web service with a unique domain name. Some tools offer custom mapping of namespaces to packages or provide refactoring of package names to resolve this problem.
- Develop a comprehensive test suite for Web Services Interoperability Organization (WS-I) conformance verification.

7.7 Summary

This chapter discusses the various concepts of integration of SOA and Web Services and the necessity and need for integration. The chapter also stresses the Web services semantics in WSDL and how carefully they must be designed before the actual implementation, and about the major roadblocks for Web services interoperability.

References

1. Aaron Skonnard. SOA: More Integration, Less Renovation. *MSDN Magazine*, http://msdn.microsoft.com/en-us/magazine/cc163850.aspx, 2010.
2. Cordys. Web Services and SOA – Enterprise Integration. http://www.cordys.com/cordyscms_com/web_services_soa.php, 2010.
3. Dave Hollander. Common Models in SOA. Progress Software, http://www.asprom.com/note/wp20.pdf, 2008.
4. Gopala Krishnan Behra. Service Integration, BPTrends. http://www.bptrends.com/publicationfiles/10%2D08%2DART%2DSOA%20and%20Integration%2DGopola%2Dfinal%2Edoc%2Epdf, 2008.
5. Kevin Francis. Workflow in Application Integration. http://msdn.microsoft.com/en-us/library/bb245667.aspx, 2010.
6. LANSA. Implementing Service Oriented Architecture Solutions (SOA). http://www.lansa.com/solutions/soa.htm, 2010.
7. Tom Laszewski. SOA and the Mainframe: Two Worlds Collide and Integrate. http://www.theserverside.com/news/1363665/SOA-and-the-Mainframe-Two-Worlds-Collide-and-Integrate, 2009.

8

Metadata Management

8.1 Evolution of Metadata Management in SOA

Metadata is data about data. More formally the string, metadata is divided into Meta and data. Meta in the Oxford Dictionary means, 'Something of a higher or second-order kind'. The word data is a data item like Birth date = '03/22/1941', but in more general sense so as to include unstructured data like text and diagrams. The scope of metadata is restricted to Information Technology. Consequently, metadata are the materialized artifacts that define the requirements for, the specifications of, design of, or even executing characteristics of an IT system, or component of that system. 'System' here is used in a very broad context. Thus, included within the scope of systems are databases, application systems, and their technology environments. Therefore, metadata is all that which is one or more levels of abstraction removed from the actual databases, applications, or their technology environments. In a computing environment, metadata would therefore include:

- Requirements.
- Functional descriptions.
- Work plans.
- Database designs through to schema DDL (data definition language).
- Application system designs possibly through to computer program source code libraries.
- Technology environment designs through to actual installation artifacts.

But within this context, would not include:

- Actual databases with data records of employees, invoices, products, and customers.
- Executing application systems.
- Operating systems and other systems software such as DBMS and Web browsers.
- Telecommunications Networks.

169

- Computers.

These are not metadata because they are 'real', while the previous list represents artifacts about the reality. But once the information system is executing, metadata may be created that describes the characteristics of the operating environment. That class of metadata would include for example:

- Computer system execution schedules.
- Computing resource consumption requirements.
- Quantity of records in particular files.
- Quantity of users by time of day for particular processes.
- Job completion and/or error messages.

8.1.1 Types of Metadata Usage

Three types of usage were identified:

- General employee information finding across repositories.
    - Dependent on metadata about people and organizations.
- Linking people to distributed business processes.
    - Provide the information to complete the process.
    - Notion of document type not easily applied.
- Content management/publishing to intranet or portal.

Over the last three decades or so, enterprise applications were built to automate a wide range of business processes. SAP is a leader in providing packaged business applications, i.e. applications for major business processes (such as Financials, Human Resources, and Customer Relationship Management) that can be customized for use by a wide variety of customers. In addition to packaged applications, enterprises also built in-house, outsourced, and otherwise acquired a wide range of applications, resulting in IT landscapes that were increasingly heterogeneous. With ever increasing connectivity within an enterprise as well as with the trading network of an enterprise, spanning their customers, suppliers, partners, and legal and administrative entities, integration across such complex IT landscapes quickly became a significant issue. Several business processes span multiple applications and the cost of integrating such systems rapidly became a limiting factor for IT. This need for integration across systems has now grown into a need for IT landscapes that are increasingly flexible, i.e. support changing of user-interface specification, and other configuration information. Design-time repositories of traditional application development tools have had a limited role and application metadata has rarely been explicitly available. Similarly, integration or EAI tools have had their own design-time repositories. In SOAs there is a need to store, manage, reason about and perform other operations on specifications of a large variety of software components, and business processes across their lifecycle.

Metadata management describes the increasing importance of metadata in today's service-oriented application landscape, and the consequent fragility inherent in architectures when faced with change. When we reach the point at which metadata drives the development and maintenance of services, evolving business requirements force us to break open and evolve our metadata first, and then to address the services dependent on that metadata. Most development methodologies and environments are not sufficiently equipped to deal with such metadata-driven change.

8.1.2 SOA and Metadata

A SOA is a metadata-driven architecture. Metadata is crucial to the development lifecycle of Web services because the long term maintainability of the SOA is at risk when the business logic expressed in services is not visible to the IT department at a higher level than in the code itself. However, there are many different kinds of metadata, not all of which are visible in application development environments. Figure 8.1 illustrates the metadata that we care about in a SOA.

The top half of this diamond represents WSDL and 'policy' metadata, which is what most developers think of when we talk about metadata in a SOA. This metadata is described in XML, hence the general understanding that services are XML based. WSDL and policy metadata are low in semantic business information and high in technical information – they provide or facilitate the plumbing that allows the services to function.

The WSDL and policy part of this equation is of low strategic value to the business because it is largely generated. It falls out of any one of a number of application development tools that might be used to design and create services, or is handcrafted according to relatively simple requirements. When change is necessary in the business logic of a SOA, developers seldom need to concern themselves with this XML – it is the visible, accessible part of the iceberg, as it were.

The lower half of this diamond describes the payload, or the messages, that the services must process in order for the business process to succeed. Payloads require a very different and altogether more fragile kind of metadata: XML schemas. Strictly speaking, services with document-centric payloads can operate very well without an external description in XML schema – that is, all payloads have an implicit schema, and there is no requirement to express the schema explicitly in XSD. Without comprehensive metadata describing the payloads, however, implementing changes to a business process quickly resembles the process of looking for a needle in a haystack.

SOA governance is ultimately the combination of policy, process and metadata. Metadata, or data about data, is the set of policies and descriptions of business services that enables discovery and appropriate usage of those services. A rich set of information about business services must be interrelated to support all of the governance and lifecycle processes, such as publication, validation and approval that are required to ensure that a SOA remains manageable. Generally speaking, there are three types of metadata: business information, technical information and governance information. Business information includes information like service type (e.g., order entry) and line of business focus (e.g., retail banking). Technical information includes transport type, authentication, interfaces and implementation. Finally, examples of governance information include the various policies and agreements discussed previously, and the relationships and dependencies between SOA elements. In a tightly-coupled world, metadata is typically defined within the code of systems and applications. SOA requires this metadata to be externalized – separated from the native system – to enable the classification and governance of these independent services. Thus, metadata becomes a key artifact that needs to be managed within a SOA.

8.1.3 Expose the Underlying Models

To repeat an earlier statement: the long-term maintainability of the SOA is at risk when the business logic expressed in services is not visible to the IT department at a higher level than in the code itself. This is an important mantra when you consider that only message-based, document-centric SOAs are likely to be successful and low cost in the long term as these allow us to rise above the point-to-point, RPC-style application integration of early service-based architectures. The latter tend over time towards time-consuming, error-prone, application-specific, and high-risk maintenance phases.

If you accept that the SOA should be message based, and your long-term goal is to achieve optimum efficiency in the development lifecycle, a best practice is to externalize the schemas, expose the models, standardize, and federate. Beyond a certain level of complexity, especially with multiple developers and teams collaborating on the development of services, the only safe way to constrain the business processes of an organization is to make the data model explicitly visible to all architects, developers, and project managers as a coherent set of XML schemas, and then to drive all service development on the basis of those schemas.

8.1.4 Advantages

The advantages of externalized schemas for message-based SOA are as follows:

- Enforceable contracts for processing behavior.
- Visible specifications for developers.
- Public interfaces for new partners in the SOA.
- Schema-based access to standard infrastructure such as parsers, transformation engines, and so on.
- Insulation for services from changes to schemas.
- Support for business analysts when planning changes.

8.1.5 Disadvantages

The disadvantages of any metadata-driven application environment are due entirely to the limitations of metadata in general and XML schemas in particular. What are these limitations? In essence, the XML schemas describing payloads are application specific, bespoke metadata that is subject to change, and requires human involvement when it evolves. Nowadays there is an expectation for change of schema. Unfortunately, schema families and their associated assets (transformations and so on) present redundancy and duplication when we try to evolve them by editing them. In any orchestrated set of Web services used and maintained by multiple development teams – for example, an order-to-invoice trading transaction involving multiple players – the externalized schemas and transformations (probably one of each per service) describe or reference the same data objects over and over again. Modifying any object presents the kind of maintenance nightmare that most of us try very hard to avoid in conventional programming environments.

8.1.6 Versioning and Impact Problems

Managing XML infrastructure is different. When developers modify schema-driven applications by modifying the schemas, two problems arise: first, the new versions of the schemas are no longer in sync with the older versions; and second, the lack of a robust, scientific mechanism for identifying where every object has been defined and referenced forces developers into a manual maintenance exercise. This is generally not a problem when there is only one schema and one developer.

The sheer proliferation of references to single objects, coupled with the number of places where objects can be reused, increases the workload of maintenance projects in an exponential curve. Typically, the only people who can carry out maintenance work beyond a certain level of complexity are highly paid system experts who become IT bottlenecks due to the level of manual work involved. Such work can be very tedious.

8.1.7 Risk

The high level of risk inherent in such a situation makes the bottom half of the diamond in Figure 8.1 equivalent to the hidden 9/10 of the iceberg, which as you will recall from the Titanic is what sinks supposedly unsinkable ships. It is a sobering thought that if you cannot manage the evolution of the schemas governing the payloads in your SOA project, there is no project available on a long run.

8.1.8 Versioning and Extensibility Issues

The first question that most organizations set out to answer at this stage is, 'How do we version schemas?' There are many different techniques applicable to the schema versioning problem, ranging from forcing instant incompatibility at one end of the spectrum (and thus forcing systems to up-grade), through to the opposite end of the spectrum where schema constraints are relaxed sufficiently to allow steadily broader ranges of content in service payloads.

These versioning techniques provide enough material for an article in their own right. A quick summary is that while it is not impossible to version schemas and the systems that depend on them, nothing comes for free, and there are very few robust mechanisms and procedures that work well.

8.1.9 Extensible Schemas

Having experienced the pain of schema versioning, the next question that organizations inevitably ask themselves is this: 'Is there a way of designing our schemas from the outset so that they are extensible?' Architects and system analysts are delighted to discover that XML Schema provides various ways of designing in extensibility, thus ensuring that schemas can be modified (read: 'extended') without affecting existing systems.

Each extension, however, has the disadvantage of making the schema considerably more complex, with the logical conclusion in the long term that your metadata reaches a level of complexity that is unmanageable. Again, the subject of schema extensibility is large enough to warrant a separate article.

8.2 Approaches to Metadata Management

In the context of SOA, metadata is largely contained within an XML schema (XSD) and the Web Services Description Language (WSDL). The XSD defines canonical (common) data representations for business objects and inherently defines data constraints such as data types and enumeration (a list of sanctioned values). With proper construction of the XSD and supporting XSD standards, metadata management for SOA can be greatly improved.

Metadata is also required for SOA to define interfaces. For Web services, the WSDL definition describes how to access a Web service and what operations it will perform. These service definitions can be discovered by service consumers through the UDDI registry. However, the UDDI specification lacks rich metadata and has no provisions for metadata governance. Packaged software such as Systinet's registry supports the UDDI standard while providing governance, services classification and a configurable taxonomy to further support services metadata.

The next step for SOA metadata management is defining the underlying meaning (semantics) of the data. The goal of providing these semantics is to automate semantic service discovery, orchestration and integration. Software vendors are recognizing the need to integrate semantics into their SOA offerings. IBM and WebMethods have acquired technology this year to add semantics to their metadata repositories.

IBM also repackaged the Meta Stage metadata repository from the Ascential acquisition into the IBM Meta Data Server with the intent to tie the product to all their Web Sphere information integration and business integration tools. Longer-term the Meta Data server will likely support the rational product line as well as the WS registry. This would present a single view of metadata across the data and SOA product lines for a powerful offering. Much of the corporate metadata is now within the data management discipline and coming from Ascential, IBM's MetaData Server has strong ties to data management.

8.2.1 Management and Enforcement of Policies over Applications

Typically in SOAs, executing a business process in an application requires integration of several loosely coupled components and services. In such an environment, the enforcement of policies in areas such as access, security or quality of service becomes a significant challenge. Especially since these policies may change frequently over time. Such policies may pertain to:

- Service operations, e.g. service performance or quality of service.
- Security, i.e. ensuring that services are called in a manner that is consistent to their specified usage and from authenticated parties.
- Management of service level agreements.
- Enforcement and management of high-level business initiatives such as regulations (e.g. the Sarbanes-Oxley act or HIPAA) and protection of intellectual property.

Current approaches towards these issues, both commercial and in research, leave a lot to be desired. For instance, a good policy management framework needs to have fairly deep knowledge of the semantics of service operations, and policy definitions, and yet needs to be able to monitor/enforce policies efficiently.

8.2.2 Metadata Management to Enable Enterprise Search

Knowledge workers in an enterprise often need to access and analyze information across various information assets in the SOA world: documents, business objects, reports, and structured data present in many different types of applications. For example, a business user might need to find information on a customer that could range from the orders placed by the customer, to the products purchased by them, the status of the orders, the financial information associated with the purchases, the service/support requests they have filed, the sales and support people the customer has interacted with and the nature of such interactions, the customer's contact information, past payment history, credit rating, and other public information on the customer, such as news

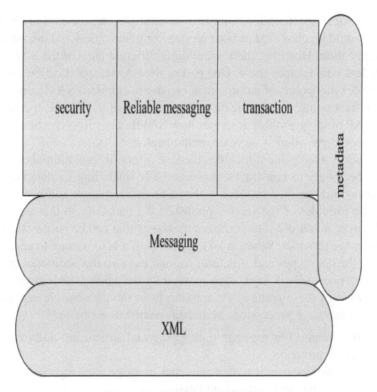

Figure 8.1 Metadata specification.

stories about the customer, product brochures, and so forth. EII technologies currently only provide basic abilities to do this. Open issues in this area include the ability to integrate retrieval across data types, performance, business processes in response to business needs. Companies such as Dell and Wal-Mart, for example, are viewed as having changed some of the fundamental business processes in their respective industries. The burden of responding to a change in the business processes of a company is one of the primary challenges posed to enterprise IT organizations. Companies also enter into new markets, or acquire/merge into other companies.

8.3 Metadata Specifications

Metadata specifications describe the structure of messages that can be sent. Some good examples are WSDL, XML Schema, and WS-Policy. Of course, there are some specifications that fit into multiple categories. For example, SAML would be classified as both service communication and infrastructure communication. However, there are entirely different parts of the SAML spec which deal with each of these. One part of the SAML specification identifies the services that points of enforcement can use to contact a SAML authority – this is infrastructure communication. Another part of the SAML specification (the SOAP binding profile) describes how SAML assertions can be sent from consumers to providers – service communication.

Metadata management includes the description information about a Web service necessary to construct a message body (including its data types and structures) and message headers so that a service requester can invoke a service. The provider of the service publishes the metadata so that a requester can discover it and use it to construct messages that can be successfully processed by the provider. When invoking a service, it is important to understand not only the data types and structures to send but also the additional qualities of service provided (if any), such as security, reliability, or transactions. If one or more of these features are missing from the message, it may prevent successful message processing. Metadata specifications include:

- *XML Schema*: For message data typing and structuring and expressing policy information.
- *WSDL*: For associating messages and message exchange patterns with service names and network addresses.
- *WS-Addressing*: For including endpoint addressing and reference properties associated with endpoints. Many of the other extended specifications require WS-Addressing support for defining endpoints and reference properties in communication patterns.
- *WS-Policy*: For associating quality of service requirements with a WSDL definition. WS-Policy is a framework that includes policy declarations for various aspects of security, transactions, and reliability.
- *WS-Metadata Exchange*: For querying and discovering metadata associated with a Web service, including the ability to fetch a WSDL file and associated WS-Policy definitions. Service binding is different for a SOA based on Web services compared to a SOA based on J2EE or CORBA, for example. Instead of binding via reference pointers or names, Web services bind using discovery of services, which may be dynamic. If the service requester can understand the WSDL and associated policy files supplied by the provider, SOAP messages can be generated dynamically to execute the provider's service. The various metadata specifications

are therefore critical to the correct operation of a SOA based on Web services.

8.3.1 Addressing

Addressing is an important requirement of extended Web services because no directory of Web services endpoint addresses exists on the Web. SOAP messages must include the endpoint address information within the message for all but the simplest MEP. WS-Addressing replaces earlier proposals called WS Routing and WS-Referral. Without an addressing solution, when you send a Web service request to a provider, typically the only address the provider has is the return address to the requester, and then only for the duration of the session. If anything goes wrong on the reply, there is no good way to retry it, basically the requester's address can be lost when there is a communication failure. Also there is no good way to specify a different return address than the requester is address. And finally, there is no way to define address schemes or to identify endpoint addresses for complicated message exchange patterns or multi-transport support.

8.3.2 Policy

Policy is necessary for expressing any extended Web services features of a service so that the requester can provide the necessary security, transaction, or reliability context in addition to the data requirements for the messages expressed in the WSDL file. Policy provides a machine-readable expression of assertions that a service requester must adhere to in order to invoke upon a provider. Does the service require security or support transactions? The latter can be very important when trying to figure out whether or not a long running, complex interaction can involve a transaction, or whether a transaction can span across all the Web services identified for it. WS-Policy is necessary for achieving interoperability for the extended features because the policy declarations are the only way in which a requester can discover whether a provider requires some or all of the extended features.

In the case of security, for example, different providers may support different kinds of tokens, such as X.509 or Kerberos. WS-Security is designed as a kind of open framework that can carry any token type. However, if the token type the provider expects is not declared, the requester can only guess at what it must be. When making the decision to invoke the provider's service, it may also be important to discover whether it supports reliability or transactions. You might want to know, for example, whether the provider's service can accept retries if the original submission fails and whether it will let you know when it has successfully received a message. Finally, you may want to know whether or not to send a transaction context to the provider to enroll the provider Web service in the transaction.

8.3.3 Acquiring Metadata

It is possible that a requester will obtain the metadata it needs using WS Metadata Exchange or another similar mechanism that queries the WSDL and associated policy files directly. WS-Metadata Exchange uses a feature of WS-Addressing called 'actions' to access the metadata published by a service provider. WS-Meta data Exchange is designed to provide any and all information about a Web service description – essentially replacing UDDI for most applications.

Developers may or may not use UDDI, despite its existence. It is fair to say that the public UDDI does not provide the metadata management facilities required to support interoperability requirements at the extended specification level and that WS-Meta data Exchange may be needed for requesters to ensure they have the information they need to achieve interoperability with providers using extended features.

8.3.4 Security

Security concerns apply at every level of the Web services specification stack and require a variety of mechanisms to guard against the numerous challenges and threats that are a part of distributed computing. The mechanisms may have to be used in combination to guard against a particular threat or combination of threats. In the Web services and SOA world, it is particularly important to evaluate the need for protection at the network layer, the message layer, and for the data in the message. Basic security protection mechanisms are built around encryption, authentication, and authorization mechanisms and typically include comprehensive logging for problem tracking [2]. The industry has achieved consensus around a single specification framework, WS-Security, although ongoing work is necessary to complete the profiles and additional related specifications. WS-Security was developed to provide message-level security on an end-to-end basis for Web services messages.

Typical HTTP-based security mechanisms, such as SSL, provide only transport-level point-to-point protection. Sometimes additional security may be provided through the use of an alternative transport mapping, such as CORBA's IIOP or Web Sphere MQ, but as with the rest of the extended features, the security specifications are written for HTTP as a kind of default or lowest common denominator transport and therefore can be applied to any transport. WS-Security headers include the ability to carry strong authentication formats such as Kerberos tickets and X.509 certificates and can use XML Encryption and XML Signature technologies for further protecting the message contents. Although a WS-Security authorization profile for the Security Assertion Markup Language (SAML) is being developed, SAML can also be used on its own for exchanging authorization information. Additional specifications from IBM, Microsoft, Verisign, and others further extend and complement WS-Security, including:

- *WS-SecurityPolicy*: Defines security assertions detailing a Web service's requirements so that the service requester can meet them.
- *WS-Trust*: Defines how to establish overall trust of the security system by acquiring any needed security tokens (such as Kerberos tickets) from trusted sources.
- *WS-SecureConversation*: Defines how to establish and maintain a persistent context for a secure session over which multiple Web service invocations might be sent without requiring expensive authentication each time.
- *WS-Federation*: Defines how to bridge multiple security domains into a federated session so that a Web service only has to be authenticated once to access Web services deployed in multiple security domains. Because Web services are XML applications, and because XML has security challenges of its own (it is basically human-readable text sent over the Internet), XML based security technologies are also often important for protecting the XML data before and after it is included in a SOAP message. These technologies include:
  - *XML Encryption*: Designed to provide confidentiality, using a variety of supported encryption algorithms, of part or all of an XML document to ensure that the contents of a document cannot be intercepted and read by unauthorized persons.
  - *XML Signature*: Designed to provide integrity, using a variety of encryption and signature mechanisms, to ensure that service providers can determine whether or not documents have been altered in transit and that they are received once and only once. XML En-

cryption and XML Signature can be used to protect Web services metadata as well as data.

8.3.5 Reliability and Messaging

Messaging includes SOAP and its various message exchange patterns (MEP). The industry has not achieved consensus on a single, unified set of specifications for advanced messaging. However, competing specifications in the categories of reliability and notification work essentially the same way, and so an amalgam of the two is used here for the sake of introduction. In general, reliable messaging is the mechanism that guarantees that one or more messages were received the appropriate number of times. Reliable messaging specifications include:

- WS-Reliability.
- WS- ReliableMessaging (from IBM and Microsoft).

Reliable messaging is designed to ensure reliable delivery of SOAP messages over potentially unreliable networks such as the HTTP-based Internet. Reliable messaging is a protocol for exchanging SOAP messages with guaranteed delivery, no duplicates, and guaranteed message ordering. Reliable messaging works by grouping messages with the same ID, assembling messages into groups based on message number, and ordering them based on sequence number. Reliable messaging automates recovery from certain transport-level error conditions that the application would otherwise have to deal with on its own. Reliable messaging also supports the concept of bridging two proprietary messaging protocols over the Internet. Finally, there is one important set of standards that defy all of these classifications: The WS-I profiles. The goal of the WS-I is to clarify existing standards, not create new ones. So, the WS-I (which is an industry consortium) defines 'profiles' that identify the pitfalls and loose-ends in existing standards – and how to avoid them. So, depending on which standard the WS-I is clarifying, a WS-I profile could be related to any of these roles.

Knowing which of these three categories a specification falls into tells you a lot about the implications of supporting it. Metadata specifications are usually used to enable automation through tools (versus manual coding). Of course, metadata without data is useless – a metadata specification always needs to fit hand-in-hand with service and/or infrastructure communication specifications. Metadata by itself (with no matching communication specification) can only serve an informational role – letting people understand information about services. UDDI has good examples of this: the taxonomies and categories within UDDI (which are service metadata) have no formal ties to impact any communication, and so this information's value is purely for people to get a better understanding of the services.

In contrast, a service communication specification without an associated metadata specification means that there will be hand coding and hand configuration. For example, if WS-Security existed, but WS-Security Policy did not, this would mean that you would have to tell the consumer what settings to use for WS-Policy, and they would have to, essentially, hard code this into their application logic or configuration. But, in either case, service communication standards impact the message flow – whether this is the content of the message, the security, or the quality of service (reliability, etc.). As an application developer, you would need to know the most about these specifications. But, to clarify, what you really need to know is how this specification impacts what you do in your application logic (for example, reliable messaging might mean you do not need to do retries for making sure important information reaches its target). It is necessary to understand the details of specification message formats, wire protocols, etc., to design the necessary infrastructure that aid in efficient communication.

8.3.6 Extended Web Services Specifications

Following the broad adoption and use of the basic Web services specifications – SOAP and WSDL – requirements have grown for the addition of extended technologies such as security, transactions, and reliability that are present in existing mission-critical applications. These extended features are sometimes also called *qualities of service* because they help implement some of the harder IT problems in the SOA environment and make Web services better suited for use in more kinds of SOA-enabled applications.

A class of applications will find the core specifications sufficient, while other applications will be blocked or hampered by the lack of one or more of the features in the extended specifications. For example, companies may not wish to publish their Web services without adequate security or may not wish to accept purchase orders without reliable messaging guarantees. The core specifications were defined with built-in extensibility points such as SOAP headers in anticipation of the need to add the extended features.

8.3.7 Standardization

Web services specifications progress toward standardization through a variety of ways, including small groups of vendors and formally chartered technical committees. As a general rule of thumb, most specifications are started by a small group of vendors working together and are later submitted to a standards body for wider adoption. Specifications initially created by Microsoft and IBM, Extended Web Services Specifications 33 together with one or more of their collaborators (these vary by specification, but typically include BEA, Intel, SAP, Tibco, and Verisign), tend to gain the most market traction.

Microsoft and IBM are the de facto leaders of the Web services specification movement and have defined or helped to define all the major specifications. Several of the WS-* specifications remain under private control at the time of writing, but we expect them to be submitted to a standards body in the near future. Standards bodies currently active in Web services include:

- *World Wide Web Consortium* (W3C): Making its initial name on progressing Web standards, notably HTTP, HTML, and XML, the W3C is home to SOAP, WSDL, WS-Choreography, WS-Addressing, WS-Policy, XML Encryption, and XML Signature.
- *Organization for the Advancement of Structured Information Standards* (OASIS): Originally started to promote interoperability across Structured Generic Markup Language (SGML1) implementations, OASIS changed its name in 1998 to reflect its new emphasis on XML. OASIS is currently home to UDDI, WS-Security, WS-BPEL, WS-Composite Application Framework, WS-Notification, WS-Reliability, Web Services Policy Language (part of the Extensible Access Control Markup Language TC), and others such as Web Services for Remote Portlets, Web Services Distributed Management, and Web Services Resource Framework.
- *Web Services Interoperability* (WS-I): Established in 2002 specifically to help ensure interoperability across Web services implementations, WS-I sponsors several working groups to try to resolve incompatibilities among Web services specifications. WS-I produces specifications called profiles that provide a common interpretation of other specifications and provides testing tools to help Web services vendors ensure conformance to WS-I specifications.

  1. Both HTML and XML are derived from SGML.

2. These specifications are not all covered in this book because the book is focused on SOA.

- *Internet Engineering Task Force* (IETF): The IETF is responsible for defining and maintaining basic Internet specifications such as TCP/IP and SSL/TLS. Their relationship to Web services is indirect in that TCP/IP is the most common communications protocol used for the HTTP transport, and basic IETF security mechanisms are used in Web services. The IETF collaborated with the W3C on XML Signature.
- *Java Community Process* (JCP): Established by Sun to promote the adoption of Java and control its evolution, the JCP is home to several Java Specification Requests (JSRs) that define various Java APIs for Web services, including JAX-RPC for the Java language bindings for SOAP, JAX-B for XML data binding, and Java APIs for WSDL.
- *Object Management Group* (OMG): Initially established to create and promote specifications for the Common Object Request Broker (CORBA), the OMG is home to specifications that define WSDL language mappings to C++ and CORBA to WSDL mappings. Web services standardization started with the submission of the SOAP 1.1 specification to the W3C in mid-2000. After that, SOAP with Attachments, XKMS, and WSDL were submitted to W3C.

At the same time, UDDI was launched in a private consortium and was later submitted to OASIS. Other major specifications submitted to OASIS include WS-Security, WS-BPEL, WS-CAF, and WS Notification [3]. More recently, WS-Addressing and WS-Policy were submitted to W3C, signaling a potential shift back toward W3C as the home of most of the major specifications. Historically, OASIS is also the home of the ebXML set of specifications, which overlap to a large extent with the Web services stack. Web Services and ebXML share SOAP, but beyond that, the stacks diverge. ebXML has its own registry and its own orchestration (or choreography) language.

8.3.8 Specification Composability

As mentioned previously, the SOAP and WSDL specifications are designed to be extended by other specifications. Two or more of the extended specifications can be combined within a single SOAP message header. For example:

```
<S:Header>
<wsse:Security>
```

```
. . .
</wsse:Security>
<wsrm:Sequence>
. . .
</wsrm:Sequence>
</S:Header>
```

This example illustrates the use of extended headers for security and reliability. The security header typically includes information such as the security token that can be used to ensure the message is from a trusted source. The reliability header typically includes information such as a message ID and sequence number to ensure the message (or set of messages) is reliably received.

Note the separate namespaces used for the security and reliability headers, wsse: and wsrm:, respectively. The headers use different namespaces so that they can be added incrementally to a SOAP message without concern over potential name clashes. Duplicate element and attribute names are not permitted in an XML document (and a SOAP message is an XML document, after all). Namespace prefixes provide a unique qualifier for XML element and attribute names, thus protecting names from duplication. This is one way in which SOAP header extensions work composably with each other. Adding extended features may or may not require modification to existing Web services – that is, the extended features can be added into the SOAP headers without changing the SOAP body in many cases. But the execution environments and mapping layers may need to change in order to handle the extensions. Certainly at least adding SOAP headers for extended features must be done within the context of knowing whether the execution environment can support them and how; otherwise, the extended headers will not work.

Web service extensions are also added to the responsibility of the SOAP processors in the execution environment. Policy declarations associated with the WSDL contracts can be used during the generation of SOAP messages to determine what should go into the headers to help the execution environment negotiate the appropriate transport protocol or to agree on features such as transaction coordination. Each additional extended feature added to the Web service invocation results in additional processing before sending the message or after receiving it. The extended features may also be related to requirements from a business process engine and may need to be supported by the registry.

8.4 Metadata Policies

Metadata Oolicies for information describing items in the repository:

1. Anyone may access the metadata free of charge.
2. The metadata may be re-used in any medium without prior permission for not-for-profit purposes provided the OAI Identifier or links to the original metadata record are given.
3. The metadata must not be re-used in any medium for commercial purposes without formal permission.

Web services are being successfully used for interoperable solutions across various industries. One of the key reasons for interest and investment in Web services is that they are well-suited to enable service-oriented systems. XML-based technologies such as SOAP, XML Schema and WSDL provide a broadly-adopted foundation on which to build interoperable Web services. The WS-Policy and WS-Policy Attachment specifications extend this foundation and offer mechanisms to represent the capabilities and requirements of Web services as Policies.

Service metadata is an expression of the visible aspects of a Web service, and consists of a mixture of machine- and human-readable languages. Machine-readable languages enable tooling. For example, tools that consume service metadata can automatically generate client code to call the service. Service metadata can describe different parts of a Web service and thus enable different levels of tooling support.

First, service metadata can describe the format of the payloads that a Web service sends and receives. Tools can use this metadata to automatically generate and validate data sent to and from a Web service. The XML Schema language is frequently used to describe the message interchange format within the SOAP message construct, i.e. to represent SOAP Body children and SOAP Header blocks.

Second, service metadata can describe the 'how' and 'where' a Web service exchanges messages, i.e. how to represent the concrete message format, what headers are used, the transmission protocol, the message exchange pattern and the list of available endpoints. The Web Services Description Language is currently the most common language for describing the 'how' and 'where' a Web service exchanges messages. WSDL has extensibility points that can be used to expand on the metadata for a Web service.

Third, service metadata can describe the capabilities and requirements of a Web service, i.e. representing whether and how a message must be secured, whether and how a message must be delivered reliably, whether a message must flow a transaction, etc. [4]. Exposing this class of metadata about the capabilities and requirements of a Web service enables tools to generate code modules for engaging these behaviors. Tools can use this metadata to check the compatibility of requestors and providers. Web Services Policy can be used to represent the capabilities and requirements of a Web service.

Web Services Policy is a machine-readable language for representing the capabilities and requirements of a Web service. These are called 'policies'. Web Services Policy offers mechanisms to represent consistent combinations of capabilities and requirements, to determine the compatibility of policies, to name and reference policies and to associate policies with Web service metadata constructs such as service, endpoint and operation. Web Services Policy is a simple language that has four elements – Policy, All, ExactlyOne and Policy Reference – and one attribute – wsp:Optional.

8.5 Metadata Exchange

The Metadata Exchange specification expands the current Web services architecture to govern the transfer of metadata, enabling endpoints to leverage other Web services specifications for secure, reliable, transacted message delivery.

Meanwhile, the update to the WS-Addressing specification enables messaging systems to support message transmission in a transport-neutral manner through networks, including networks with processing nodes, such as endpoint managers, firewalls and gateways, Microsoft officials said. The changes reflect community feedback and experience gained by vendors in implementing the technology and from interoperability workshops, company officials said. The changes include an improved definition-of-request reply and a new WSDL (Web Services Description Language) binding and fault codes.

'WS-MetadataExchange – as well as WS-Addressing – is a key step on the WS roadmap,' said Jason Bloomberg, an analyst with ZapThink LLC, in Waltham, Mass.

> WS-MetadataExchange is a key part of enabling the standards-based exchange of metadata. We like to say that metadata is the lifeblood of service-oriented architecture [SOA], because the core layer of abstraction behind SOAs – what provides flexible services and hides the complexity of heterogeneous back-ends – runs on metadata. These announcements also show that IBM and Microsoft

are still working together to further standardization efforts across the industry.

8.6 Metadata Challenges in SOA

Metadata Challenges in SOA adoption can be classified and put into three categories. These three categories are People, Process, and Technology.

8.6.1 Challenges with People

When adopting SOA, a buy-in and awareness should be there at all levels in the organization. Real benefits of SOA can be realized only in the middle or long-term. It will be unrealistic to expect real benefits in the short-term. Having said that, SOA adoption should have good buy-in and commitment from the senior management team that is funding it. The CXO team should be willing to wait and continue to show the interest and commitment in SOA.

SOA training is an important thing. Individual departments and project teams should have enough knowledge and commitment to SOA for the initiative to be successful. While the commitment from CXO is required for SOA adoption to continue, the commitment from individual departments and project teams is required to develop the applications using SOA principles. With SOA, new roles, new products, and new processes are introduced. At the core of the SOA solution are the services that get reused. As organizations mature in their SOA adoption, they will involve new roles, such as service librarian. The service library will have a complete catalog of the services, and the service librarian plays an important role in taking care of the issue of service proliferation and the redundancy of services in the organization. People will need to learn these new roles. Current processes will be updated with these new roles, and new processes will be introduced. People in the organization should start following these new processes. Generally, people are comfortable working in silos as they have complete control when working independently. SOA brings in a big shift towards sharing culture and working in teams. The complexity increases as more teams get involved and more collaboration is needed. Planning becomes more complex than the traditional application development.

8.6.2 Challenges in Adopting New Processes

As stated earlier, among the challenges for people, SOA adoption will bring in additional roles and processes. This section describes the challenges with respect to some of the key processes in SOA adoption.

8.6.3 SOA Governance

Governance is about establishing and communicating the policies that employees must follow. It is about giving employees the tools they need to be compliant with the policies. It is about providing visibility into the levels of compliance in the organization. It is about mitigating the deviations. SOA governance extends corporate governance, IT governance, and architecture governance. In other words, SOA governance can be as good as the architecture governance; architecture governance can be as good as IT governance; and IT governance can be as good as corporate governance.

Some of the key challenges with respect to SOA governance are

- ROI for SOA.
- SOA Investment & Funding.
- Metadata Management.
- Reuse Promotion.

8.6.4 ROI for SOA

The EA teams in large organizations generally have good buy-in and are able to get the required funding to start SOA initiatives. But there are a good number of organizations that are asked to build a business case and develop an ROI model for SOA adoption. These organizations in most cases tend to forget to measure the real benefits after SOA implementation is done. Some of the issues that are hindering development and measurement of ROI are:

- Historical data to make the estimates and to quantify reuse.
- Differentiating the benefits from SOA vs. non-SOA.

SOA business value can primarily be classified in two buckets. The first one is increased revenue, and the second one is reduced costs. With increased revenue (for example, higher sales, improved time to market, improved customer satisfaction, reduced risks), it is difficult to measure and pin down the benefits to SOA. However, when we look at cost savings, IT cost savings (service reuse, hardware/software licenses, etc) is relatively easy to measure and can be pinned to SOA.

8.6.5 SOA Investment and Funding

Traditionally, departments allocate funds for IT projects that cater to their specific requirements. Typical funding in an enterprise is usually project-based. With SOA architecture, the all-pervasive nature of SOA architecture makes it difficult to find one single business sponsor to recognize and pay for the long-term goals that it would achieve. When starting SOA adoption, initially it could be funded by mid level managers' discretionary budgets, but as the adoption grows, higher levels of management need to become involved. Enterprise level SOA requires funding obtained from the highest level, which could be the CIO or CFO. The SOA library of components should be regarded as the line of business administered and funded separately. Several organizations are already having or going to have separately funded enterprise-wide architecture groups to govern and architect shared SOA solutions.

8.6.6 Metadata Management

It is the reuse and metadata-driven development that make SOA solutions flexible. Examples of metadata are the service contracts that abstract the underlying implementation, the business rules that are not hard coded into the application code, the policies that are required to be enforced at right policy enforcement points, and the composite applications that are built by orchestrating the business services. All of these elements are called metadata. It is the data about the data. Managing this metadata is a challenge in SOA adoption, particularly when the SOA adoption increases in the organization. Registry and repository solutions will be key players in simplifying or automating the metadata management-related processes and artifacts. However, it also requires manual processes in addition to the use of tools for metadata management.

8.6.7 Reuse Promotion

Promoting reuse is a key to the success of SOA adoption. Without reuse, the benefits of SOA are not realized. With no proper reuse promotion, organizations will end up with the same old silos of solutions with no reuse. Reuse becomes after the fact requirement and will not be considered or valued during the development time. Organizations will end up having a proliferation of services with the same set of services implemented again and again. One of the key things in promoting reuse is to make the service information available in the organization. Create awareness in the project teams about the available services, and, more importantly, reward for building and utilizing services. These processes need to be put in place in the organization to promote reuse and to realize better benefits of SOA.

8.6.8 Challenges with Technology

SOA is not something new. SOA concepts have been there for a long time now. Advancements in Web services and WS-* standards are key in SOA gaining momentum. New products like ESB, Process Orchestration, WSM, XML gateways, and Registries & Repositories are available as foundation for SOA implementation. Some of these products are still maturing with the addition of new features and by new ways of complying with the standards. Interoperability standards for Web services are improving. Dealing with these maturing standards and products is a challenge to the SOA teams. In order to avoid vendor lock-in, teams should try to avoid using proprietary features offered by product vendors. People have to learn these new technologies and products. They will have to learn new products across the development lifecycle. Teams should learn new products for building services, and for testing and deploying them. They will have challenges with doing capacity and sizing exercises for the solutions based on these new product stacks. The team will have challenges in doing QoS testing for this new set of products.

8.7 Summary

Metadata management and its role on building a Web service application is elaborated with discussion on appropriate and potential technical challenges in maintaining the metadata. This chapter also summarizes governance concept of SOA and what are the future challenges in the technological growth of SOA.

References

1. Asir S. Vedamuthu. Microsoft Corporation Daniel Roth, Microsoft Corporation, Understanding Web Services Policy, msdn, 2010.
2. M. Gudgin et al. Web Services Addressing 1.0 – Core. http://www.w3.org/TR/2006/REC-ws-addr-core-20060509/, 2006.
3. K. Ballinger et al. Web Services Metadata Exchange (WS-Metadata Exchange). http://schemas.xmlsoap.org/ws/2004/09/mex/, 2004.
4. Della-Libera et al. Web Services Security Policy Language (WS-SecurityPolicy). http://schemas.xmlsoap.org/ws/2005/07/securitypolicy, 2005.

9

Security Issues

9.1 Building Blocks of Security

Considering the 'big-picture' of SOA Security, it may be important to understand different aspects of security, role of AAA (Authentication, Authorization and Auditing) in SOA Security, encryption, industry standard specifications etc. This section introduces the key goals of SOA security, nature of security threats and finally some practical SOA security implementations in the industry. Considering Web service is the most widely accepted approach to implement SOA, the technical aspects in this section focus more on Web service security.

9.1.1 Goals of SOA Security

As the adoption of SOA increases, the boundary of services 'grows' beyond internal applications. To achieve true re-usability, it may be required for organizations to expose services to third-parties, partners or even end-customers over insecure networks such as the Internet. Services are organizational assets and exposing them without appropriate security considerations poses a big threat to the organization in the form of un-authorized access, misuse of services, overuse of services and also hacker vulnerability.

To address the above risks, the main goals of SOA security are:

- *Authentication* – Allowing access only to the intended application that invokes the service. In traditional security approaches, this is the standard 'credential'-based security such as a login ID/password pair, certificates, etc.
- *Authorization* – Controlling access only to a defined set of services and/or operations within a service. This is the classical 'role'-based security to restrict access to a subset of functions.

- *Auditing/Monitoring* – Recording all invocations of a service to address the 5 Ws of security – Who, What, When, Where & Why. This is crucial to identify an attack and trace the attacker. Also, auditing constitutes a 'digital' record of all activities within the SOA infrastructure.
- *Federation* – When a service requires authentication against another external system, federation is used. Federation is an extension of authentication that helps the service provider to establish trust between the provider's security domain and an external domain. So the external provider 'trusts' the request and considers it authenticated without expecting an additional credential.
- *Integrity* – Goal of integrity checking is to ensure that the XML data entering in the form of a Web service request is not corrupted.
- *Policy* – The idea of a policy in SOA security is the capability of the service provider to specify Web service's conditions under which the service is provided. For example, the condition may require that the request to a Web service be encrypted.
- *Throttling* – It is a concept to control the 'bandwidth' offered by a service. Though not directly related to security, throttling is typically used to protect the service infrastructure so that service consumers do not 'overuse' the services. In some cases, this can also be used to prevent 'denial of service' attacks.
- *Confidentiality/Network Level Security* – The goal of network level security is to encrypt data packets transmitted to and from the SOA infrastructure. This is to prevent any packet-sniffing tools to intercept any passwords.
- *Hack-proof* – Even if a genuine service consumer successfully authenticates and has necessary role permissions on a service, it is very important to ensure that service boundaries are not crossed to prevent several Web-service specific attacks such as XPath injection, XML structure manipulation, schema attacks, etc.

9.1.2 SOA Security Implementation – A Logical View

The diagram in Figure 9.1 depicts the building blocks of a typical SOA security implementation.

The SOA Access Gateway is a critical component that enforces security to the SOA platform. In addition to the standard firewall based security, the SOA Access Gateway can specifically interpret Web Service requests. The firewall is used to allow 'port' and 'IP-Level' access. Like a Firewall, the SOA Access

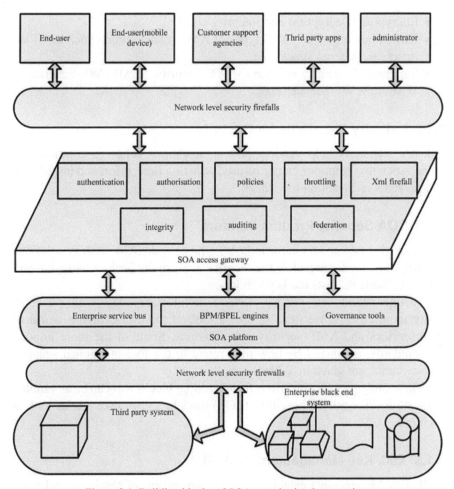

Figure 9.1 Building blocks of SOA security implementation.

Gateway is a hardware box that directly fits into the IP network. However, the SOA Access Gateway is a step ahead and provides the following functions:

- Authentication – In the form of WS-Security tokens.
- Authorization – Acts as a policy enforcement point (PEP) and policy definition point (PDP).
- Auditing – Captures usage statistics.
- Throttling – Allows restricting bandwidth for a particular service. Example Service A can be invoked only at 80 transactions per second, whereas Service B can be invoked at 100 transactions per second.

- Encryption/Decryption and Integrity checks.
- XML Firewall – Detects all types of XML related threats such as XPath injection, Schema attacks, etc.
- Supports all security standards – WS-Security, SAML, WS-Federation, WSPolicy, WS-Metadata, etc.

9.1.3 Industry Standards for Security

Industry standards help vendors and organizations follow a common approach such that solutions can be re-used reducing time, effort and investment and prevents re-inventing the wheel.

9.1.4 SOA Security Product Vendors

There are several vendors in the industry who provide SOA security solutions that help organizations realize the security goals in the form of hardware and software. Table 9.2 lists the key vendors.

To understand how service-oriented architectures deal with these security requirements, we need to take a look at the ever-evolving world of Web services and XML security specifications. Some of the more important standards, organized by how they relate to the fore-mentioned security requirements, are given in Table 9.3.

These and other specifications form building blocks that can be assembled to create service-oriented security (SOS) models. Let us take a brief look at each one.

9.1.5 XML Key Management (XKMS)

XML Key Management establishes a standard means of obtaining and managing public keys. Even though XKMS is compatible with a number of public key infrastructure (PKI) technologies, it does not require any of them, and removes the need for integrating proprietary PKI products.

The XML Key Management specification consists of two complementary standards: the XML Key Registration Service and the XML Key Information Service specifications. Together, they allow for the integration of a number of security technologies, including digital signatures, certificates, and revocation status checking [1]. For instance, XKMS can enlist XML-Digital Signatures to protect the integrity of XML document content.

Table 9.1 Security standards.

SOA Security Goal	Standards	Overview
Authentication	WS-Security	Originally drafted by IBM, Microsoft and VeriSign, WS-Security defines a standard way of specifying username and encrypted password in SOAP headers
	WS-Trust	WS-Trust aims to enhance the WS-Security by providing additional features such as a Security Token Service (STS). STS offers services such as Token Exchange, Token Issuance and Validation. This standard is approved by OASIS.
	WS-Secure Conversation	WS-Secure Conversation is another extension to WS-Security which defines the means to create a security context and allows a series of message exchanges (conversation) to be done when authentication
Authorization	XACML	XACML (eXtensible Access Control Markup Language) is an XML schema specification to define authorization and entitlement policies. XACML addresses the lack of fine-grained access control granularity in SAML.
Federation	SAML	SAML (Security Assertion Markup Language) is primarily an XML-based standard authentication language to authenticate across different security domains, such as SSO – Single Sign-On.
Policy	WS-Policy	WS-Policy is a standard way for service providers to specify a wide range of service requirements (policies) such as maximum message size, service traffic handling capacity, etc. This standard is approved by OASIS.
	WS-Security Policy	WS-Security Policy standard defines security related policies based on WS-Policy and WS-Secure Conversation standard. This standard is approved by OASIS.
	WS-Metadata Exchange	WS-Metadata Exchange specification defines a mechanism for service clients to retrieve service metadata information such as Schema, WSDL and WS-Policy.
Encryption/ Confidentiality	XML-Encrypt	XML-Encrypt is a W3C recommendation to encrypt sensitive fields within XML documents and also to specify the encryption algorithm that is used.
	XML-Signature (also known as XML-DSig)	XML-Signature is also a W3C recommendation for XML digital signature processing to allow clients to digitally sign an XML. This ensure message integrity, which allows service providers to detect content corruption, malicious content, etc. Advanced versions of XML-DSig already exists such as XAdES (XML Advanced Electronic Signatures).
	XKMS	XML Key Management System is a W3C recommendation which allows developers to secure communications using public key infrastructure (PKI). The specification describes protocols for distributing and registering public keys to be used in conjunction with XML-Encrypt and XML-Signature. XKMS consists of two parts – XKISS – XML Key Information Specification & XKRSS – XML Key Registration Service Specification.
	SSL	Needless to say, Secure Sockets Layer is the basic foundation technology to ensure transport-level security.

Table 9.2 Vendors of security products.

Sno	Product Name	Vendor
1	IBM Datapower (Access Gateway – Hardware)	IBM
2	Cisco ACE XML Gateway (Hardware)	Cisco
3	Intel XML Security Gatewy (Hardware)	Intel
4	Web Services – Domain Boundary Controller (Hardware)	Xtradyna
5	Amberpoint SOA Management System (Software)	Amberpoint

Table 9.3 Security requirements.

Identification	WS-Security Framework
Authentication	Extensible Access Control Markup Language (XACML)
Authorization	Extensible Rights Markup Language (XrML) XML Key Management (XKMS) Security Assertion Markup Language (SAML) .NET Passport
Confidentiality	WS-Security Framework XML-Encryption Secure Sockets Layer (SSL)
Integrity	WS-Security Framework XML-Digital Signatures

9.1.6 Extensible Access Control Markup Language (XACML) and Extensible Rights Markup Language (XrML)

The XACML specification consists of two related vocabularies: one for access control and one that defines a vocabulary for request and response exchanges. Through these languages, the creation of fine-grained security policies is possible.

It is important not to confuse XACML with the WS-Policy specification, which also can be used to define policies, and is considered part of the WS-Security framework. An additional specification that may be relevant to your environment, if you transport files with different digital formats, is the Extensible Rights Markup Language (XrML).

9.1.7 Security Assertion Markup Language (SAML) and .NET Passport

Single sign-on technologies help address an administration problem that has emerged when an enterprise environment consists of applications that independently control user access lists. If a single sign-on system is not already in place, adding Web services can contribute to the decentralized proliferation of user credentials. By opening up new integration channels, more users may be required to access applications. This can lead to an ever-increasing maintenance effort.

Popular technologies for single sign-on include the Security Assertion Markup Language (SAML) and the .NET Passport. SAML provides mechanisms for both authentication and authorization processes. Both request and response message formats are defined to facilitate the transmission of necessary credentials within a Web service activity. Microsoft's .NET Passport is a competing technology, and relies on proprietary protocols for handling authentication. It also introduces a centralized management system for user credentials, which differs from SAML's decentralized approach. Interoperability options between the two technologies do exist, and continue to improve

9.1.8 XML-Encryption and XML-Digital Signatures

These two key specifications protect the actual content within XML documents.

The XML-Encryption specification contains a standard model for encrypting both binary and textual data, as well as a means of communicating information essential for recipients to decrypt the contents of received messages.

XML-Digital Signatures establishes a standardized format for representing digital signature data. Digital signatures establish credibility within a message, as they assure the recipient that the message was in fact transmitted by the expected partner service. It also provides a means of communicating that the message contents were not altered in transit, as well as support for standard non-repudiation. As with the XML-Encryption standard, ML-Digital Signature also supports binary and textual data.

9.2 Challenges, Threats and Solutions

Software designers and developers are being challenged to build efficient security measures into their project work as computing is increasingly distributed via Web application services and Service-Oriented Architectures (SOA).

Research recently conducted by the Ponemon Institute and CipherOptics found that only 12% of IT professionals surveyed believed that cyber-crime threats were lessening in severity.

Among the findings analyzed in their report 'Network Security 2.0', the practice of sending data in clear text over third-party networks, the increasing presence of organized crime, growing complexity of networks, devices and applications, and the desire to enforce and easily manage network encryption were cited as prevalent threats to network security.

There are significant risks to organizations, banks and financial service companies specifically due to inadequate security and the employee's using peer-to-peer (P2P) network services.

9.2.1 A Rich Vein for Cyber Crime

Ponemon and Cipher Optics were surprised to find that only 35% of respondents said 'no' when asked if their network environments permit sensitive or confidential information to pass over third-party networks in clear, readable text – clearly an incentive for organizations to make use of Layer 2 data encryption technologies such as those offered by Cipher Optics and others.

Using Tiversa technology, Dyne and his colleagues collected and categorized tens of thousands of P2P files and searches related to the top 30 US banks.

'For one bank, we found a spreadsheet with 23,000 business accounts including their contact names and addresses, account numbers, company positions, and relationship managers at the bank', the authors noted in the report. Dynes recounted

> We set out to see what kind of documents were out there that might result in economic damage to banks. We were able to look at both the documents and at the searches people were doing and we found there are people out there searching for things that other people would not want them to see, like credit card numbers, bank statements, Experian reports – and there are documents out there that banks would not want others to see.

Private and confidential information is being exposed by banks, their law firms and other service suppliers, right on through to and including the landscaping companies they use, according to Dynes:

> We came across documents such as how to put a computer on a network at the bank. We saw one document at a law firm that talked about the merger between two financial institutions; we saw one, a proposal to banks for various services. Some of these could be quite damaging and they would be found by searches that folks are using.

9.2.2 Higher Performance or More Security

Security threats are an ever-present risk regardless of IT infrastructure – client-server, SaaS or SOA, noted Sandy Carter, vice president of SOA and WebSphere strategy, channels and marketing at IBM.

Carter told TechNewsWorld:

> The most prevalent security challenges today are those presented by malware, including computer hacking, consumer identity theft crimes, viruses, or other security vulnerabilities motivated by illegal profits. The emergence of organized cyber crime, coupled with the necessity for organizations to comply with myriad government regulations that ensure security and privacy of client data has made securing IT infrastructures through industry standard technologies critical for organizations of all sizes and in all industries.

Management and IT staff face the vexing problem of striking the right balance between realizing the intended benefits of SOA and Web application services while at the same time securing information and their IT infrastructures.

Carter continued

> As a company establishes its SOA strategy it must carefully take into account its security requirements to ensure there aren't any trade-offs while gaining flexibility, reuse, cost reductions and productivity gains. Today, there are many SOA-specific products designed to address the security concerns associated with extending a SOA to different parts of the organization, the Web and to customers and partners alike.
>
> Products such as the IBM Data Power SOA were created to bolster the security in a SOA. These SOA appliances offer a unique way to simplify deployments, improve performance and enhance the security of SOA implementations. Web Sphere Data Power SOA appliances feature critical transformation, acceleration, security and routing functions to help ease the deployment of SOA implementations.

Most IT organizations have standardized on Ethernet as their LAN technology and the WAN vendors are moving to managed Ethernet services, such as metro Ethernet and MPLS/VPLS offerings, said Scott Palmquist, CipherOptics' senior vice president of product management.

9.3 Layer Level Security

As the complexity and sophistication of application and business logic within Web services increases, so does the risk associated with putting a corporation's business intelligence 'out there'. An increased level of service-oriented application functionality leads to more integration opportunities with external business partners. While the security framework established by the many specifications that provide standards for XML and Web services is relatively new, most of the principles behind these standards are not. The fundamental characteristics of a primitive security architecture are just as relevant to service-oriented environments as they are to traditional distributed applications.

The established concepts are:

- *Identification*: The recipient of a message needs to be able to identify the sender.
- *Authentication*: The recipient of a message needs to verify that the claimed identity of the sender is valid.
- *Authorization*: The recipient of a message needs to determine the level of the sender's security clearance. This can relate to which operations or which data the sender is granted access to.
- *Integrity*: A message remains unaltered during transmission, up until actual delivery.
- *Confidentiality*: The contents of a message cannot be viewed while in transit, except by authorized services.

A common response to addressing WS-Security requirements is: 'we'll just encrypt it with SSL'. Many assume that this is a valid security measure, because they associate the Internet-based communication between Web services with the traditional interaction between a browser and a Web server.

The Secure Sockets Layer (SSL) technology enables transport-level security. Accessing a secured site using a browser is a fairly safe procedure, when communication is encrypted. That is because the connection is exclusive to the client browser and the Web server that acts as the gateway to internally hosted application logic.

Once a Web service transmits a message it does not always know where that message will be traveling to, and how it will be processed before it reaches its intended destination. A number of intermediary services may be involved in the message path, and a transport-level security technology will protect the privacy of the message contents only while it is being transmitted between these intermediaries.

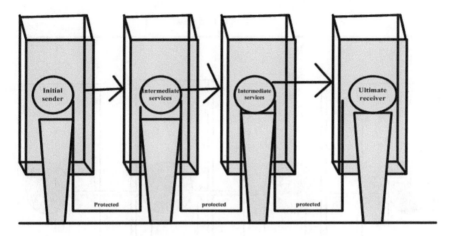

NOT PROTECTED BY TRANSPORT _ LEVEL SECURITY

Figure 9.2 SSL layer message transport.

In other words, SSL cannot protect a message during the time that it is being processed by an intermediary service. This does not make SSL unnecessary within a service-oriented communications framework, it only limits its role. A number of additional technologies (discussed throughout this section) are required to facilitate the message-level security required for end-to-end protection.

9.3.1 The WS-Security Framework

This important document establishes fundamental and conceptual security standards, and also defines a set of supplementary specifications that collectively form a Web service-centric security framework. WS-Security (also known as the Web Services Security Language) can be used to bridge gaps between disparate security models, but also goes beyond traditional transport-level security to provide a standard end-to-end security model for SOAP messages.

Because service-oriented environments sometimes require that intermediaries be involved with the delivery of messages, an end-to-end security model is needed, so that the contents of SOAP messages remain protected throughout a message path. This is different from the traditional point-to-point model, for which transport-level security has generally been sufficient.

End to End

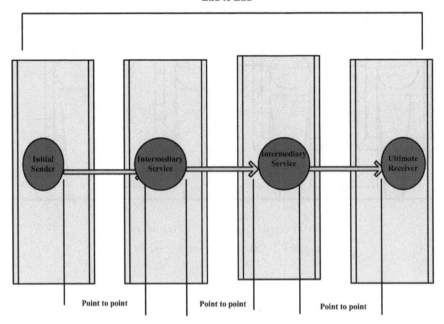

Figure 9.3 Service-oriented environment.

Table 9.4 WS-Security framework.

WS-Policy	WS-Trust	WS-Privacy
WS-SecureConversation	WS-Federation	WS-Authorization

In fact, you could view a message path involving intermediaries as a series of point-to-point connections.

To secure a message path from end-to-end, WS-Security implements security measures through the use of SOAP header blocks that travel with the message. Although the WS-Security framework complements a number of the specifications described earlier in this section, it is supported further by a series of supplementary standards (see Table 9.4).

The first layer below the framework is frequently referred to as the Policy Layer, as it provides a number of building blocks for the creation of trust and privacy policies. The next row, known as the Federation Layer, builds on these policies to unify disparate trust domains.

9.3.1.1 WS-Policy

Though not limited to providing security-related policies, this standard is a key part of the WS-Security framework. Existing corporate security polices can be expressed through policy assertions that can then be applied to groups of services.

9.3.1.2 WS-Trust

This specification establishes a standard trust model used to unite existing trust models, so that the validity of exchanged security tokens can be verified [2]. WS-Trust provides a communications process for requesting the involvement of third-party trust authorities to assist with this verification

9.3.1.3 WS-Privacy

Organizations can use WS-Privacy to communicate their privacy policies and check to see whether requestors intend to comply to these policies. WS-Privacy works in conjunction with WS-Policy and WS-Trust.

9.3.1.4 WS-Secure Conversation

Various security models can be supplemented with WS-Secure Conversation, which establishes a standard mechanism for exchanging security information between Web services. It provides formal definitions for the creation and exchange of security contexts and associated session keys.

9.3.1.5 WS-Federation

There are numerous ways of integrating different trust domains (or realms) when utilizing the WS-Security, WS-Policy, and WS-Trust standards. The WS-Federation specification provides a series of standards and security models for achieving a federation – an environment where a level of trust has been established between disparate trust domains.

9.3.1.6 WS-Authorization

WS-Authorization provides a standard for managing information used for authorization and access policies. As part of this standard, the manner in which claims are represented within security tokens is established.

9.3.2 An Example

Due to the vast diversity of security specifications, we do not have the luxury of delving into the concepts and language syntax of each of the standards we have discussed so far. To give you a glimpse into what a SOAP header block containing security information looks like, though, we have provided a brief example that demonstrates two fundamental parts of the WS-Security framework.

When a service requestor makes a request of a service provider, it asserts a claim regarding its security clearance. It is then up to the service provider to validate this claim. A service requestor may provide a number of claims in order to communicate different aspects of its security status. This set of claims is contained within a security token.

Security tokens can signed with a signing authority, allowing them to be further verified by the recipient of the message containing the token. Unsigned security tokens often consist of login credentials supplied by the service requestor.

The following parent Security construct establishes a WS-Security header block that consists of a token with login information:

```
<Security ...>
   <UsernameToken>
     <Username>
       Terl
     </Username>
     <Password>
       onedaytherewillonlybeonesecurityspec
     </Password>
   </UsernameToken>
</Security>
```

The UsernameToken construct contains Username and Password elements that represent the credentials claim.

9.4 Message Level Security

One simple way to secure the PO Web service would be to provide legitimate users with a token. This approach assumes that each SOAP message containing a PO document sent by a consumer will contain a username/password pair, which is checked against the database when it arrives at the service provider.

While this approach allows the service provider to obtain the information about the consumer to make an authorization decision, a significant disadvantage is that the credentials passed within the message are actually

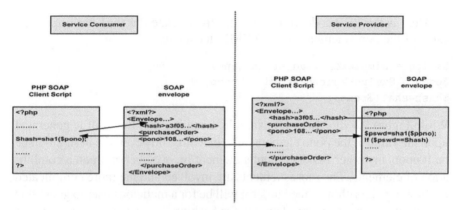

Figure 9.4 Security mechanism.

independent of the message payload, and thus, once obtained by a malicious user, may be used to consume the service on behalf of a legitimate user. Of course, you can still use SSL to ensure transport-level security. Often, though, a SOAP message sent from a service consumer to a service provider is processed by an intermediate service or services, running the risk of a malicious user stealing the password travelling with the message.

This section discusses how you might work around the above issue by using a hash generated from the value of a particular element or elements of the PO document being transmitted with the message, rather than sending a fixed token. On the client, you might include that hash as part of the SOAP message payload also containing the PO document as the other part. The server in turn is responsible for retrieving the hashed token from the message and checking whether this hash corresponds to the PO that arrived in the same message [3]. Depending on the algorithm used to generate a hash, each new PO document may come with a potentially different hashed token, which makes it harder for a malicious user to illegally access the service.

It is important to realize that the above security mechanism does not ensure a private way to transfer the data, since the payload of a SOAP message being transmitted is not encrypted. As for data integrity, you may be fully confident that the message has not been modified in transit only if the hash transmitted within the message was generated upon the entire payload rather than some parts of it. The security mechanisms prevents unauthorized users from consuming the services.

The security mechanism discussed here might look like the diagram in Figure 9.4.

Here is the PHP code you might use to generate the sha1 hash upon the value of the pono element of a PO XML document:

```
$xmlpo = simplexml_load_string($po);
$pono = $xmlpo->purchaseOrder->pono;
$hash=sha1($pono);
```

While Figure 9.4 illustrates an example assuming that the hash is generated upon the value of the pono element only, in reality, however, the hash might be built upon any other PO document's element or, even better, upon a combination of elements. The more elements are involved and the more complicated the hashing algorithm is, the harder it will be for a malicious user to guess that algorithm. When choosing elements for hashing, it is always a good idea to consider the element whose value uniquely identifies the document, since it will be most likely used as the primary key when storing the document in the database. So utilizing the pono in this particular example is essential, since an attempt to submit a new PO document with the same pono will fail due to the primary key constraints placed upon the database table holding incoming PO documents.

As you can see, with the above approach, you do not even need to create and hold the security accounts in the database, since the security measures are incorporated in a SOAP message itself, thus enabling message-level security.

> As an alternative to including credentials in the SOAP message body, you might include them in the SOAP message headers. To achieve this with the PHP SOAP extension, you might use the following predefined classes: SoapHeader and SoapVar. Using SOAP message headers to send secure messages, as well as implementing WS-Security authentication, is discussed in detail later, in the Using SOAP Message Headers and Using WS-Security for Message-Level Security sections.

The first step is to create the WSDL document for the updated PO Web service. For that, you might create the following document and save it as *po_secure.wsdl* in the *WebServices\wsdl* directory:

```
<?xml version="1.0" encoding="utf-8"?>
<definitions name ="poServiceSecure"
    xmlns:http="http://schemas.xmlsoap.org/wsdl/http/"
    xmlns:soap="http://schemas.xmlsoap.org/wsdl/soap/"
    xmlns:xsd="http://www.w3.org/2001/XMLSchema"
    xmlns="http://schemas.xmlsoap.org/wsdl/"
```

```
    targetNamespace="http://localhost/WebServices/wsdl/
    po_secure.wsdl">
    <message name="getPlaceOrderInput">
        <part name="hash" element="xsd:string"/>
            <part name="po" element="xsd:string"/>
    </message>
    <message name="getPlaceOrderOutput">
        <part name="body" element="xsd:string"/>
    </message>
    <portType name="poServiceSecurePortType">
        <operation name="placeOrder">
          <input message="tns:getPlaceOrderInput"/>
          <output message="tns:getPlaceOrderOutput"/>
        </operation>
    </portType>
    <binding name="poServiceSecureBinding"
            type="tns:poServiceSecurePortType">
        <soap:binding style="rpc"
            transport="http://schemas.xmlsoap.org/soap/http"/>
        <operation name="placeOrder">
          <soap:operation
          soapAction="http://localhost/WebServices/ch4/
          placeOrder
      <input>
              <soap:body use="literal"/>
          </input>
          <output>
              <soap:body use="literal"/>
          </output>
        </operation>
    </binding>
    <service name="poServiceSecure">
        <port name="poServiceSecurePort"
            binding="tns:poServiceSecureBinding">
          <soap:address
  location="http://localhost/WebServices/ch4/SoapServer_
  secure.php"/>
        </port>
    </service>
</definitions>
```

As you can see, this WSDL assumes two message parts in the input message of the *placeOrder* operation defined here. So, make sure that the binding style defined in the document is rpc.

Next, you might want to create the PHP handler class representing the underlying logic of the service discussed here. For that, you might create the following purchaseOrder.php script in the *WebServices\ch4* directory:

```
$this->checkOrder($hash, $pswd);
    $sql = "INSERT INTO purchaseOrders VALUES(:po)";
    $query = oci_parse($conn, $sql);
    oci_bind_by_name($query, ':po', $po);
    if (!oci_execute($query)) {
        throw new SoapFault("Server","Failed to insert PO");
    };
    $msg='<rsltMsg>PO inserted!</rsltMsg>';
    return $msg;
    }
    private function checkOrder($hash, $pswd) {
    if ($pswd!=$hash) {
        throw new SoapFault("Server","You're not authorized to
                            consume this service");
    }
    }
  }
?>
```

In this *purchaseOrder* class, in the first highlighted code block you load the PO XML document as a *SimpleXMLElement* object and then extract the value of the pono element. Next, you generate the *sha1* hash upon the extracted *pono* and call the *checkOrder* private method of *purchaseOrder*. The code for this method is also highlighted in the listing and is used to check to see whether the hash generated here is equal to the hash that arrived with the message. If there is a mismatch, a *SoapFault* exception is thrown.

The implementation of the SOAP server script to be used in this example is straightforward. It is assumed that you save the following SOAP server script as *SoapServer_secure.php* in the *WebServices\ch4* directory:

```
<?php
//File: SoapServer_secure.php
require_once "purchaseOrder_secure.php";
$wsdl= "http://localhost/WebServices/wsdl/po_secure.wsdl";
$srv= new SoapServer($wsdl);
```

```php
$srv->setClass("purchaseOrder");
$srv->handle();
?>
```

Now that you have the service created, all that is left is to build a client script to test the newly created service. For that, you might create the *SoapClient_secure.php* client script shown below:

```php
<?php
//File: SoapClient_secure.php
$wsdl = "http://localhost/WebServices/wsdl/po_secure.wsdl";
$xmldoc = simplexml_load_file('purchaseOrder.xml');
$pono = $xmldoc->pono;
$hash=sha1($pono);
$podoc=$xmldoc->asXML();
$client = new SoapClient($wsdl);
try {
  print $result=$client->placeOrder($hash, $podoc);
}
catch (SoapFault $exp) {
  print $exp->getMessage();
}
?>
```

> Before you can execute this SOAP client script, you need to have a purchaseOrder.xml document containing a PO XML document. So, make sure to copy the purchaseOrder.xml document from the *WebServices\ch2* to *WebServices\ch4* directory.

In this client script, you extract the value of the pono element from the PO loaded as a SimpleXMLObject from purchaseOrder.xml, and then generate an sha1 hash upon the extracted value. The generated hash and the PO document converted into a string are then passed to the placeOrder SOAP function as arguments.

Unlike the above client, the following one uses a more complicated algorithm for generating the hash. In particular, it generates an sha1 hash upon the pono and shipName concatenated together.

```php
<?php
//File: __SoapClient_secure.php
$wsdl = "http://localhost/WebServices/wsdl/po_secure.wsdl";
$xmldoc = simplexml_load_file('purchaseOrder.xml');
```

```php
$pono = $xmldoc->pono;
$shipName = $xmldoc->shipTo->name;
$mix=$pono.$shipName;
$hash=sha1($mix);
$podoc=$xmldoc->asXML();
$client = new SoapClient($wsdl);
try {
 print $result=$client->placeOrder($hash, $podoc);
}
catch (SoapFault $exp) {
 print $exp->getMessage();
}
?>
```

Now if you try to run the *_SoapClient_secure.php* script shown above, you should get the following error message:

You're not authorized to consume this service

This is because you have not changed the authentication algorithm on the server side yet. To handle this task, you might modify the *purchase-Order_secure.php* script as shown overleaf (the script has been cut down to save space):

```php
<?php
//File _purchaseOrder_secure.php
 class purchaseOrder {
   public function placeOrder($hash, $po) {
     ...
     $xmlpo = simplexml_load_string($po);
     $pono = $xmlpo->pono;
     $shipName = $xmlpo->shipTo->name;
     $mix=$pono.$shipName;
     $pswd=sha1($mix);
     $this->checkOrder($hash, $pswd);
     ...
   }
 ...
 }
?>
```

Now the mechanism of generating the hash on the client agrees with the one used on the server. So, if you now run the *_SoapClient_secure.php* script, you should not have a problem.

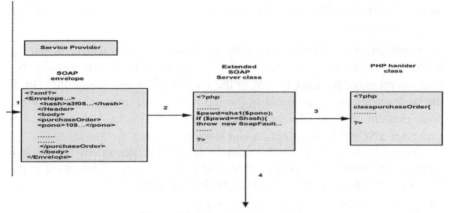

Figure 9.5 Secure message arrival.

9.4.1 Using SOAP Message Headers

In the preceding section you saw an example of how you might implement message-level security by including a hash generated upon the data from the document being transmitted into the message. The mechanism discussed there assumed that you pass the hash generated as part of the message payload, which means you had to define another part of the input message in the WSDL document in order to carry the hash. However, sending the hash as part of the payload is probably not a good idea, because the hash, acting as a security measure in this case, can be considered metadata rather than user data transmitted within the SOAP message payload.

Figure 9.5 gives a graphical depiction of the process that takes place on the server side when a secure message arrives.

Here is the explanation of the steps in the figure:

Step 1: The service provider receives the message containing a PO document as the payload and the hash as a header.

Step 2: The overridden handle method of an extended SOAP server class checks whether the hash that arrived with the message as a header is equal to the hash generated upon the *pono* element of the PO document composing the message payload.

Step 3: If the check performed in Step 2 returns true, the SOAP server passes the PO document to an instance of the PHP handler class for further processing.

Step 4: Otherwise, the server stops processing the message, throwing a SOAP exception.

An important point about the security mechanism discussed here is that the SOAP server processes the hash passed in as a header of the message before the message payload is sent to the handler class for processing. So, if the hash passed in is not equal to the hash generated upon the value of the *pono* element of the PO document that arrived as the payload, then the server generates a SOAP fault exception and stops processing the message.

As usual, let us start with creating the WSDL document for the updated poServiceSecure. For that, create the *po_headers.wsdl* document in the *WebServices\ch4* directory, which might look as follows:

```
<?xml version="1.0" encoding="utf-8"?>
<definitions name ="poServiceSecure"
        xmlns:http="http://schemas.xmlsoap.org/wsdl/http/"
        xmlns:soap="http://schemas.xmlsoap.org/wsdl/soap/"
        xmlns:xsd="http://www.w3.org/2001/XMLSchema"
        xmlns="http://schemas.xmlsoap.org/wsdl/"
        targetNamespace=
          "http://localhost/WebServices/wsdl/po_headers.wsdl">
    <message name="getPlaceOrderInput">
        <part name="po" element="xsd:string"/>
    </message>
    <message name="getPlaceOrderOutput">
        <part name="body" element="xsd:string"/>
    </message>
    <portType name="poServiceSecurePortType">
        <operation name="placeOrder">
            <input message="tns:getPlaceOrderInput"/>
            <output message="tns:getPlaceOrderOutput"/>
        </operation>
    </portType>
    <binding name="poServiceSecureBinding"
            type="tns:poServiceSecurePortType">
        <soap:binding style="document"
            transport="http://schemas.xmlsoap.org/soap/http"/>
        <operation name="placeOrder">
          <soap:operation
          soapAction="http://localhost/WebServices/ch4/
            placeOrder"/>
          <input>
              <soap:body use="literal"/>
```

```
          </input>
          <output>
              <soap:body use="literal"/>
          </output>
       </operation>
    </binding>
    <service name="poServiceSecure">
       <port name="poServiceSecurePort"
                    binding="tns:poServiceSecureBinding">
          <soap:address
 location="http://localhost/WebServices/ch4/
          SoapServer_headers.php"/>
       </port>
    </service>
</definitions>
```

There are a couple of points worth noting about the WSDL document shown opposite. First, the input message of the *placeOrder* operation defined here consists of one part only – you do not need to define the *hash* part any more, since you are not going to transmit a hash as part of the message payload. Second, you may use the binding style document, since the input message payload is going to contain only a PO document.

```php
<?php
 //File purchaseOrder_headers.php
 class purchaseOrder {
   public function placeOrder($po) {
    if(!$conn = oci_connect('xmlusr', 'xmlusr',
       '//localhost/XE')){
        throw new SoapFault("Server", "Failed to connect
        to database");
    };
    $sql = "INSERT INTO purchaseOrders VALUES(:po)";
    $query = oci_parse($conn, $sql);
    oci_bind_by_name($query, ':po', $po);
    if (!oci_execute($query)) {
        throw new SoapFault("Server","Failed to insert PO");
    };
    $msg='<rsltMsg>PO inserted!</rsltMsg>';
    return $msg;
   }
 }
?>
```

The *secSoapServer* class is shown in the following snippet that extends the *SoapServer* predefined SOAP extension class, overriding the parent's *handle* method. It is assumed that you save this class as *secSoapServer.php* in the *WebServices\ch4* directory.

```php
<?php
 //File: secSoapServer.php
 class secSoapServer extends SoapServer {
  function handle($req) {
   $env = simplexml_load_string($req);
   $hash= $env->xpath('//ns1:hash');
   $hash = (string) $hash[0];
   $po= $env->xpath('//po');
   $po = simplexml_load_string((string)$po[0]);
   $pono = $po->xpath('//pono');
   $pono = (string)$pono[0];
   $pswd=sha1($pono);
   if ($pswd!=$hash) {
        throw new SoapFault("Server","You're not authorized to
                            consume this service");
   };
   parent::handle();
  }
 }
?>
```

In the overridden *handle* method, you first convert the value of the argument passed in to the method and representing the request message received by the server into a *SimpleXMLElement* object, which makes it possible for you to access the request message as XML. In particular, you use the *xpath SimpleXMLElement* method to access the hash encapsulated within the message header block. Using the same method, you obtain the message payload, specifying *//po* as the path argument for the xpath method. If you recall from the *po_headers.wsdl* document discussed earlier in this section, *po* is the name of the input message part that represents the message payload. Next, you load the obtained payload as another *SimpleXMLElement* object that you then use to access the *pono* element in the PO document representing the payload. It is explained later in this section why you have to create another *SimpleXMLElement* object to access the payload, rather than accessing it via the *SimpleXMLElement* object created earlier and representing the entire message. Finally, to make use of the parent *handle* method functionality, you explicitly call this method.

Now that you have the *secSoapServer* class created, you can put it into action with the following SOAP server script, which you should save as *SoapServer_headers.php* in the *WebServices\ch4* directory:

```php
<?php
//File: SoapServer_headers.php
require_once "purchaseOrder_headers.php";
require_once "secSoapServer.php";
$wsdl= "http://localhost/WebServices/wsdl/po_headers.wsdl";
$srv= new secSoapServer($wsdl);
$srv->setClass("purchaseOrder");
$srv->handle($HTTP_RAW_POST_DATA);
?>
```

To test the newly created service, you might create and then execute the following client script:

```php
<?php
//File: SoapClient_headers.php
$wsdl = "http://localhost/WebServices/wsdl/po_headers.wsdl";
$xmldoc = simplexml_load_file('purchaseOrder.xml');
$pono = $xmldoc->pono;
$hash=sha1($pono);
$header = new SOAPHeader('http://localhost/WebServices/ch4/
                         headers', 'hash', $hash);
$client = new SoapClient($wsdl);
$client->__setSOAPHeaders($header);
$podoc=$xmldoc->asXML();
try {
  print $result=$client->placeOrder($podoc);
}
catch (SoapFault $exp) {
  print $exp->getMessage();
}
?>
```

In this particular example, the header block transports the hash used as a security measure.

Now let us turn back to the *secSoapServer* class, discussed a bit earlier in this section. Examining the *handle* method overridden in this class, you may wonder why you would want to use another *SimpleXMLElement* object to access the payload, despite the fact that you already have a *SimpleXMLElement* object representing the entire message. To understand why you have to do it this way, it would be a good idea to look at the request message passed to the *handle* method for processing.

There are several ways in which you can do this. For example, you might make use of the _getLastRequest method of a *SoapClient* on the client side. As a result, you should get the following message:

```
<SOAP-ENV:Envelope ...>
  <SOAP-ENV:Header>
   <ns1:hash>da30b3a3056d477be870db86a140a4a36cf7b243</ns1:hash>
  </SOAP-ENV:Header>
  <SOAP-ENV:Body>
   <po>
    &lt;?xml version="1.0"?&gt;
    &lt;purchaseOrder&gt;
     &lt;pono&gt;108128476&lt;/pono&gt;
     &lt;billTo&gt;
      ...
    &lt;/purchaseOrder&gt;
   </po>
  </SOAP-ENV:Body>
</SOAP-ENV:Envelope>
```

So, to extract the hash from the above message you use the following code in the *handle* method:

```
$env = simplexml_load_string($pack);
$hash= $env->xpath('//ns1:hash');
$hash = (string) $hash[0];
```

However, since the PO document composing the message payload contains HTML entities, obtaining the value of the *pono* element of the PO is a bit tricky. First, you obtain the string representing the PO and containing HTML entities. Next, you load this string as a *SimpleXMLElement* object and then get the *pono* element with the *xpath* method as follows:

```
$po= $env->xpath('//po');
$po = simplexml_load_string((string)$po[0]);
$pono = $po->xpath('//pono');
$pono = (string)$pono[0];
```

This is not the case, though, if the message payload is defined as XML rather than a string. For example, you might use the following WSDL document to describe the *poServiceSecure* service discussed here:

```
<definitions name ="poServiceSecure"
        xmlns:http="http://schemas.xmlsoap.org/wsdl/http/"
        xmlns:soap="http://schemas.xmlsoap.org/wsdl/soap/"
        xmlns:xsd="http://www.w3.org/2001/XMLSchema"
        targetNamespace=
            "http://localhost/WebServices/wsdl/po_headers.wsdl"
        xmlns:xsd1="http://localhost/WebServices/schema/po/"
        xmlns="http://schemas.xmlsoap.org/wsdl/">
    <import namespace="http://localhost/WebServices/schema/po/"
            location="http://localhost/WebServices/schema/po.xsd"
  />
    <message name="getPlaceOrderInput">
        <part name="po" element="xsd1:purchaseOrder"/>
    </message>
...
</definitions>
```

In this case, the request message issued by a consumer of *poServiceSecure* would look as follows:

```
<SOAP-ENV:Envelope ...>
  <SOAP-ENV:Header>
   <ns1:hash>da30b3a3056d477be870db86a140a4a36cf7b243</ns1:hash>
  </SOAP-ENV:Header>
  <SOAP-ENV:Body>
   <SOAP-ENV:placeOrder>
    <po>
     <pono>108128476</pono>
     <shipTo>
      ...
    </po>
   </SOAP-ENV:placeOrder>
  </SOAP-ENV:Body>
</SOAP-ENV:Envelope>
```

This simplifies things at the server end. Now, the code extracting the hash and the value of the *pono* element in the overridden handle method of *secSoapServer* might look as follows:

```
$env = simplexml_load_string($pack);
$hash= $env->xpath('//ns1:hash');
$hash = (string) $hash[0];
$pono= $env->xpath('//pono');
$pono = (string)$pono[0];
```

In the above example, you use only one *SimpleXMLElement* object, loading the entire request message into it and then extracting first the hash and then the value of the pono element of the PO document composing the message payload.

9.4.2 Using WS-Security for Message-Level Security

While the security approach discussed in the preceding section may be efficient in many PHP SOAP extension-based solutions and is easy to maintain, it does not represent a standard security mechanism.

If you want to employ a standard SOAP security mechanism, consider WS-Security, a core security specification describing a mechanism for implementing message-level security, providing the means of encapsulating security measures in SOAP messages.

Actually, the PHP SOAP extension provides no support for WS-Security. To take advantage of the technology, you have to explicitly create the required WS-Security headers and put them into the SOAP message. The header of the message in this case should look as follows:

```
<SOAP-ENV:Header>
 <ns1:Security
        xmlns:ns1="http://schemas.xmlsoap.org/ws/2003/06/secext">
  <ns1:UsernameToken>
   <ns1:Username>yourusername</ns1:Username>
   <ns1:Password>yourpassword</ns1:Password>
  </ns1:UsernameToken>
 </ns1:Security>
</SOAP-ENV:Header>
```

> It is interesting to note that WS-Security is not the only WS-* specification that utilizes message headers – virtually all WS-* specifications do that. For example, WS-Addressing transports message exchange information with SOAP headers.

Now that you know what a WS-Security header looks like, let us create a client script that would be able to post messages containing such a header.

To start with, you need to create two classes that will be used in the process of creating a WS-Security header. The first class should look as follows:

```
<?php
 //File: UsernameToken.php
```

```php
class UsernameToken {
 private $Username;
 private $Password;
  public function __construct($Username, $Password) {
    $this->Username = $Username;
    $this->Password = $Password;
  }
 }
?>
```

When creating an instance of this class you will have to specify the username and password to be encapsulated in the WS-Security header being created.

The second class should look as follows:

```php
<?php
 //File: varUsernameToken.php
 class varUsernameToken {
  private $UsernameToken;
  public function __construct($UsernameToken) {
    $this->UsernameToken = $UsernameToken;
  }
 }
?>
```

This class will be used to create a *SoapVar* variable from an instance of the *UsernameToken* class discussed previously.

Assuming that you have saved the above classes in the *Username-Token.php* and *varUsernameToken.php* scripts respectively, you can now create the following client script:

```php
<?php
 //File: SoapClient_wssecurity.php
 require_once 'UsernameToken.php';
 require_once 'varUsernameToken.php';
 //setting up the variables
 $ns1 = 'http://schemas.xmlsoap.org/ws/2003/06/secext';
 $wsdl = "http://localhost/WebServices/wsdl/po_headers.wsdl";
 //generating the hash
 $xmldoc = simplexml_load_file('purchaseOrder.xml');
 $pono = $xmldoc->pono;
 $hash=sha1($pono);
 //building WS-Security tags
 $usr = new SoapVar('usr', XSD_STRING,null,null,null,$ns1);
```

```
$pswd = new SoapVar('pswd', XSD_STRING,null,null,null,$ns1);
$tok = new UsernameToken($usr, $pswd);
$token = new SoapVar($tok, SOAP_ENC_OBJECT,null,null,
'UsernameToken',$ns1);
$varToken = new varUsernameToken($token);
$token = new SoapVar($varToken, SOAP_ENC_OBJECT,null,null,
'UsernameToken',$ns1);
$header = new SOAPHeader($ns1, 'Security', $token);
//creating the client
$client = new SoapClient($wsdl, array('trace' => 1));
$client->__setSOAPHeaders($header);
$podoc=$xmldoc->asXML();
try {
 print $result=$client->placeOrder($podoc);
}
catch (SoapFault $exp) {
 print $exp->getMessage();
}
print "REQUEST:\n".$client->__getLastRequest()."\n";
?>
```

You define a *SoapVar* object upon an instance of the *UsernameToken* class defined earlier, and then use this *SoapVar* object when defining an instance of the *varUsernameToken* class. The latter in turn is used when defining another *SoapVar* object, which is then passed to the *SoapHeader* constructor.

For simplicity, this client script uses the same WSDL document as in the preceding example. This means that when you execute the above client, the *SoapServer_headers.php* server script discussed in the preceding section will be invoked. So, the authentication will fail, because the secSoapServer class utilized within *SoapServer_headers.php* is not supposed to work with a WS-Security header. In a real-world scenario, though, it is assumed that the receiver understands WS-Security.

However, the purpose of the client script discussed here is to show how you can set up a WS-Security header rather than how you can handle it on the server side. To achieve this goal, you create a *SoapClient* in debugging mode, and then use the *_getLastRequest* method to print out the request message. In this case, the request message will be printed regardless of whether the server fails to process the message or not.

To complete this example, though, you might create the SOAP server that will understand WS-Security headers. For that, you might want to create a class that extends the *SoapServer* class, similar to *secSoapServer* used in the preceding example, in order to handle WS-Security headers on the server side.

9.5 Data Level Security

The fundamental security specifications for protecting Web services data are XML Signature and XML Encryption.

XML Signature (also called XMLDsig, XML-DSig, XML-Sig) defines an XML syntax for digital signatures and is defined in the W3C recommendation XML Signature Syntax and Processing. Functionally, it has much in common with PKCS#7 but is more extensible and geared towards signing XML documents. It is used by various Web technologies such as SOAP, SAML, and others.

XML signatures can be used to sign data – a resource – of any type, typically XML documents, but anything that is accessible via a URL can be signed. An XML signature used to sign a resource outside its containing XML document is called a detached signature; if it is used to sign some part of its containing document, it is called an *enveloped* signature; if it contains the signed data within itself it is called an *enveloping* signature.

Structure
An XML Signature consists of a `Signature` element in the `http://www.w3.org/2000/09/xmldsig#` namespace. The basic structure is as follows:

```
<Signature>
  <SignedInfo>
    <SignatureMethod />
    <CanonicalizationMethod />
    <Reference>
       <Transforms>
       <DigestMethod>
       <DigestValue>
    </Reference>
    <Reference /> etc.
  </SignedInfo>
  <SignatureValue />
  <KeyInfo />
  <Object />
</Signature>
```

The `SignedInfo` element contains or references the signed data and specifies what algorithms are used. The `SignatureMethod` and `CanonicalizationMethod` elements are used by the `SignatureValue` element and are included in `SignedInfo` to protect them from tampering. One or more `Reference` elements specify the resource being signed by URI reference and any transforms to be applied to the resource prior to signing. A transformation can be a XPath-expression that selects a defined subset of the document tree.

`DigestMethod` specifies the hash algorithm before applying the hash. `DigestValue` contains the result of applying the hash algorithm to the transformed resource(s).

- The `SignatureValue` element contains the Base64 encoded signature result – the signature generated with the parameters specified in the `SignatureMethod` element – of the `SignedInfo` element after applying the algorithm specified by the `CanonicalizationMethod`.
- The `KeyInfo` element optionally allows the signer to provide recipients with the key that validates the signature, usually in the form of one or more X.509 digital certificates. The relying party must identify the key from context if `KeyInfo` is not present.
- The `Object` element (optional) contains the signed data if this is an *enveloping signature*.

9.5.1 Validation and Security Considerations

When validating an XML Signature, a procedure called $\hat{C}$ore Validation is followed.

1. *Reference Validation*: Each `Reference`'s digest is verified by retrieving the corresponding resource and applying any transforms and then the specified digest method to it. The result is compared to the recorded `DigestValue`; if they do not match, validation fails.
2. *Signature Validation*: The `SignedInfo` element is serialized using the canonicalization method specified in `CanonicalizationMethod`, the key data is retrieved using `KeyInfo` or by other means, and the signature is verified using the method specified in `SignatureMethod`.

This procedure establishes whether the resources were really signed by the alleged party. However, because of the extensibility of the canonicalization and transform methods, the verifying party must also make sure that what was actually signed or digested is really what was present in the original data, in other words, that the algorithms used there can be trusted not to change the meaning of the signed data.

9.5.2 XML Canonicalization

The creation of XML Signatures is a bit more complex than the creation of an ordinary digital signature because a given XML Document (an 'Infoset', in common usage among XML developers) may have more than one legal serialized representation. For example, whitespace inside an XML Element is not syntactically significant, so that <Elem> is syntactically identical to <Elem>.

Since the digital signature is created by using an asymmetric key algorithm (typically RSA) to encrypt the results of running the serialized XML document through a Cryptographic hash function (typically SHA1), a single-byte difference would cause the digital signature to vary.

Moreover, if an XML document is transferred from computer to computer, the line terminator may be changed from CR to LF to CR LF, etc. A program that digests and validates an XML document may later render the XML document in a different way, e.g. adding excess space between attribute definitions with an element definition, or using relative (vs. absolute) URLs, or by reordering namespace definitions. Canonical XML is especially important when an XML Signature refers to a remote document, which may be rendered in time-varying ways by an errant remote server.

To avoid these problems and guarantee that logically-identical XML documents give identical digital signatures, an XML canonicalization transform (frequently abbreviated to *C14n*) is employed when signing XML documents (for signing the SignedInfo, a canonicalization is mandatory). These algorithms guarantee that logically-identical documents produce exactly identical serialized representations.

Another complication arises because of the way that the default canonicalization algorithm handles namespace declarations; frequently a signed XML document needs to be embedded in another document; in this case the original canonicalization algorithm will not yield the same result as if the document is treated alone. For this reason, the so-called *Exclusive Canonicalization*, which serializes XML namespace declarations independently of the surrounding XML, was created.

9.5.3 Benefits

XML DSig is more flexible than other forms of digital signatures such as Pretty Good Privacy and Cryptographic Message Syntax, because it does not operate on binary data, but on the XML Infoset, allowing to work on subsets of the data, having various ways to bind the signature and signed information, and perform transformations [4]. Another core concept is canonicalization, that is to sign only the 'essence', eliminating meaningless differences like whitespace and line endings.

XML Encryption, also known as XML-Enc, is a specification, governed by a W3C recommendation that defines how to encrypt the contents of an XML element.

Although XML Encryption can be used to encrypt any kind of data, it is nonetheless known as 'XML Encryption' because an XML element (either an EncryptedData or EncryptedKey element) contains or refers to the cipher text, keying information, and algorithms.

Both XML Signature and XML Encryption use the KeyInfo element, which appears as the child of a SignedInfo, EncryptedData, or EncryptedKey element and provides information to a recipient about what keying material to use in validating a signature or decrypting encrypted data.

The KeyInfo element is optional: it can be attached in the message, or be delivered through a secure channel.

These specifications, along with SAML, XACML, and XKMS, are not specific to Web services because they are general to XML and are not specifically adapted to SOAP and WSDL XML Signature defines how to verify that the contents of an XML document have not been tampered with and arrived unchanged from the way they were sent. XML Encryption describes how to encrypt all or part of any XML document so that only the intended recipient can understand it. It is especially important to consider using these XML security technologies when the XML data needs to be protected outside the context of a SOAP message and when the Web services metadata needs to be protected from unauthorized access.

WS-Security uses XML Signature and XML Encryption to help ensure confidentiality and integrity of SOAP messages, but it does not describe how to use these XML technologies outside the context of SOAP and WSDL, which may be important for some applications, especially those storing XML in a kind of intermediate format between transmissions.

9.6 Summary

The Service-Oriented Architecture (SOA) concept is now embraced by many companies worldwide. However, because of its loosely coupled nature and its use of open access (HTTP), SOA adds a new set of requirements to the security landscape. This chapter highlights on the security aspects of SOA and how to implement security constraints to provide authentication, confidentiality and message integrity. The chapter also highlights on setting security at different levels like data level security, message level security, etc.

References

1. Torry Haris. Migration and Security in SOA. University of Leeds, http://specs.xmlsoap.org/ws/2004/09/mex/WSMetadataExchange, 2004.
2. Thomas Eri. *An Overview of the WS-Security Framework*. SOA Systems Inc., 2009.
3. Yuli Vasili. SOA: Implementing Message – Level Security. .NET BPEL Microsoft SOA Web Services, 2010.
4. Web Services Security. *What's Required to Secure a Service-Oriented Architecture*. An Oracle White Paper, 2008..

10

Quality of Services in Enterprise Application Integration

10.1 Introduction to Enterprise Application Integration (EAI)

Today Information Technology (IT) has become very critical for successful functioning of any enterprise in this world. Every form of business thrives on some form of automation. Earlier, the automation had been custom developed, but nowadays everything seems to be through packaged applications, which in a way reduced the amount of software development significantly. Unfortunately these packaged applications, which are self-contained, have become stove-pipe applications. The requirements for next-generation software systems mandate the integration of these stove-pipes with new forms of business logic. The term Enterprise Application Integration (EAI) has become a recent entrant into the jargon of the active software industry. EAI is a buzzword that represents the task of integration of various applications so that they may share information and processes freely. Thus EAI is the creation of robust and elegant business solutions by combining applications using common middleware and other viable technologies. With these realizations, EAI was created by industry analysts to help information technology organizations to understand the emergence of a type of software that eases the pains of integration. EAI is the nexus of technology, method, philosophy and desire to finally address years of architectural neglect.

10.1.1 Why EAI?

In the past, enterprise system architectures have been poorly planned. Many organizations built systems based on the cool technology of the day without realizing how these systems would somehow, someday, share information. There are a number of organizations fitted with different types of open and

229

proprietary systems each with its own development, database, networking and operating system. This resulted in a heterogeneous environment. Overcoming these ill-effects there came a number of ad-hoc technologies and methods. But nothing seemed to be a perfect solution considering the complexity associated with these systems. EAI has come as a boon for enterprise architects to set right everything, which today's enterprises face. EAI has become a sophisticated set of procedures with newly refined technologies, such as middlewares and message brokers that allow users to tie systems together using a common glue code. There are powerful tools and techniques to perform EAI successfully.

Another important factor that is driving enterprises toward the promised land of EAI is the broad acceptance of packaged applications, such as Enterprise Resource Planning (ERP) applications. These, called as stove-pipe applications that address and solve very narrow problems within departments, has ruled the enterprises for a long time and this is the time that package vendors and enterprises have started to realize the importance of applications integration to face the daunting tasks ahead. It has been found that EAI has the wherewithal to link many disparate systems including ERP applications.

As organizations begin to realize the necessity of interconnection of disparate systems in order to meet the needs of the business, the significance of integration technology is being simultaneously felt. Thus it becomes a very important milestone for every enterprise to effectively architect, design, and develops systems based on EAI technology. As there are a number of excellent EAI technologies, it needs a steep learning curve to acquire the knowledge and wisdom to identify when to apply EAI technology, selecting appropriate technology, architect a solution, and they applying it successfully to the problem.

10.1.2 Types of EAI

An enterprise system is made up of business processes and data. So when an IT expert contemplates to use EAI technology, he has to first understand how these business processes are automated and the importance of all business processes. This understanding will bring out a lot of useful hints for determining the amount of work needed, how much time it will take, which business processes and data are to be integrated etc. Apart from this initial and first task of exploration, the primary knowledge needed is at what level, the integration process has to be performed in an enterprise application as there are mainly four levels, such as data level, application interface level, method level, and user interface level in an application.

1. Data-level EAI is the process and the techniques and technology of transferring data between data stores. This can be described as extracting information from one database, if need, processing that information, and updating the same in another database [1]. The advantage of data-level EAI is the cost of using this approach. Because there may not be any changes in the application code and hence there is no need for testing and deploying the application resulting in a little expenditure. Also the technologies providing mechanisms to move data between databases, as well as reformats that information are relatively inexpensive considering the other EAI levels and their applicable enabling technologies.

2. Application interface level EAI refers to the leveraging of interfaces exposed by custom or packaged applications. Developers make use of these interfaces to access both business processes and simple information. Using these interfaces, developers are able to bring many applications together, allowing them to share business logic and information. The only limitations that developers face are the specific features and functions of the application interfaces.

   This particular type of EAI is most applicable for ERP applications, such as SAP, PeopleSoft and Bann, which will expose interfaces into their processes and data, do so in very different ways. The most preferred EAI technology for this type is message brokers as these can extract the information from one application, put it in a format understandable by the target application and transmit the information.

3. Method level EAI is the sharing of the business logic that may exist within the enterprise. Applications can access methods on any other application. The mechanisms to share methods among applications are many including distributed objects, application servers, and TP Monitors. An ORB can take the call of one application to methods stored in other applications. An application server can be a shared physical server for a shared set of application servers. Most of the integration has been happening at this level as there are a number of robust technologies to accomplish this type.

4. User interface level EAI is a more primitive approach. Architects and developers are able to bundle applications by using their user interfaces as a common point of integration. For example, mainframe applications that do not have database or business process-level access may be ac-

cessed through the user interface of the application. This type is not a preferred one even though on many occasions, this is the only way of approaching integration task.

10.1.3 EAI Technologies

As seen above, integration task can be accomplished successfully at different levels of an application. Thus there are different compact and elegant technologies to fulfill these goals. There are various middlewares, Object Request Brokers, message brokers, Web technology like XML etc. It becomes mandatory for an architect to have a solid understanding of these technologies, the merits and demerits of each technology, where each shines and is flexible and scalable before embarking on the grand task of integration. There are point-to-point middleware, such as remote procedure calls (RPCs) and message-oriented middleware (MOM), database-oriented middleware, and transactional middleware including TP monitors and application servers. Also distributed object computing facilitated by Object Request Brokers and message brokers are available now for successful EAI.

There is a big hype about the newly crowned Web technology XML that can do wonders as the standard integration mechanism. XML provides a common data exchange format, encapsulating both metadata and data. This permits applications and databases to exchange information without having to understand anything about each other. The primary factor in using XML technology for EAI is that XML is portable data. Finally the Java factor in EAI is also making waves among the EAI architects.

Java, a revolutionary method for building Web-born applications, is now maturing enough to be of benefit to the enterprise and to EAI. Java has become a solid technology for coding enterprise-scale mission-critical applications. Java RMI is already the Java version of distributed object computing architecture and in the recent past, Sun Microsystems has come out with a enterprise component model, referred to as EJB and there came a number of robust EJB-compliant application server. As Java provides portable code, it is bound to play a very important role in uniting enterprise applications.

JMS for EAI – JMS facilitates EAI. It takes two applications to work in concert in order to carry out some new form of transaction. JMS and XML combine well to support client/server technologies that had until recently dominated the integration landscape, marking a healthy trend toward open, flexible infrastructures as well as promoting reusable integration strategies that can be applied to a wide variety of integration projects.

Today's integration products frequently employ something of a messaging infrastructure, along with data transformation and migration tools. Java Message Service (JMS) APIs provide implementation-independent interfaces for message-oriented middleware products, such as Progress SonicMQ and IBM MQSeries. These enterprise messaging systems allow stove-pipe applications to communicate events and to exchange data while remaining physically independent. Data and events can be exchanged in the form of messages via topics or queues, which provide an abstraction that decouples participating applications. Thus there is a growing number of activities in making Java-based enterprise technologies and services as the able ally to EAI.

10.1.4 The Business Advantages of EAI

There are quite a number of distinct benefits being accrued by business houses from EAI. Here comes a brief of what a middleware-enabled EAI can do to the organizations.

Middleware-enabled EAI – Enterprise Application Integration is the creation of new strategic business solutions by combining the functionality of an enterprise's existing applications, commercial packaged applications, and new code using a common middleware. Middleware refers to technology that provides application-independent services that mediate between applications. Middleware also represents the software products that implement these middleware services. There were mechanisms before the arrival of these middleware technologies, such as CORBA, for integrating applications in enterprises. But these mechanisms were found inflexible, requiring very high effort and highly complex.

But middleware has brought some spectacular benefits for enterprises to integrate their applications without much complexity. Middleware is a software tool. Middleware provides elegant and easy mechanisms by which applications can package functionality so that their capabilities are accessible as services to other applications. Middleware is able to hide the complexities of the source and target systems, thereby freeing developers from focusing on low-level APIs and network protocols and allowing them to concentrate on sharing information. Information in different enterprises has not been organized and formatted in the same manner. Thus the information to be shared among applications in different places has to go through some translations and conversions as it flows from one application to another. These capabilities are being provided by present-day middleware technologies. Finally the middleware comes with mechanisms that help applications to coordinate business processes.

EAI is attractive for developing new applications because few changes to existing legacy or packaged applications are needed and because there is no necessity for extensive programming or custom interfaces. EAI can make use of existing application programming interfaces (APIs) and databases. Suppose there is no APIs exist, EAI may access an application's functionality by mimicking normal users through the application's user interface, using screen-scraping technology. Screen scrapping is the copying of data from specific locations on character-based screens that are displayed on an end-user computer.

The ultimate goal of EAI is to allow an organization to integrate diverse applications quickly and easily. By employing EAI technologies effectively, an enterprise can leverage its existing investments to provide new and advanced products and easy to use services, to improve its relationships with customers, suppliers and other stake-holders and to streamline its operations. EAI makes it possible for the enterprise to greatly simplify interactions with other enterprise applications by adopting a standard approach to integration in a long term perspective. Further, once an EAI infrastructure has been put in place, new EAI-based applications can usually go online more quickly than traditionally developed applications because an enhanced technical infrastructure exists on which to base future development. Thus EAI plays a very critical role in making an enterprise competitive.

10.1.5 EAI for Critical New Solutions

Improving Customer Relationships – Customers perceive that there are a number of departments or line of business in an enterprise. Customers often want to be treated by enterprises, which they do business with, as a royal quest. When dealing with one department, they do not want to be required to provide information that they have already supplied to another department. To overcome such kinds of problems, an enterprise seeks the help of EAI technologies, which can make the customers to feel the difference in the relationships with enterprises. Sometimes the enterprise wants to be able to take advantage of all that it knows about a customer. Knowing the products that a customer previously has bought can create opportunities for selling other products or additional services related to previous transactions. Also it does not matter at all if the customer sometimes interacts with the enterprise via the Web, or via a call center or in some occasions in person. That is, all the information should be integrated. Achieving improved customer relationships demands application integration. The information relevant for a customer should be available in an integrated form, even though that information may be scattered in numerous stove-pipe applications developed to support various lines of business.

Improving Supply-Chain Relationships – Apart from the customers, enterprises also want to improve relationship with supply-chain partners and other outside organizations through EAI technologies. There are a plenty of opportunities available for electronic information exchange. Sharing information, enterprises open new vista for effective coordination. Partners can also leverage new technologies to create new services. By establishing electronic links with its shipping partners, a retailer, for example, can offer enhanced order status tracking.

Improving customer relationships and acquiring higher levels of integration with supply-chain partners arises new security concerns. If the Internet is being used as a communication channel, rather than dial-up or dedicated lines, an enterprise must take necessary steps to ensure the confidentiality of information flow. Controls must be in its place to ensure that a partner can not see information relevant to other partners.

Integration requirements for interactions with partners are similar to those for interaction with customers. New applications often will be required to integrate with several existing stove-pipe applications. The ability to support information exchange by making use of technology such as eXtensible Markup Language (XML) is becoming a key factor in automated business-to-business integration. A variety of front-end channels may be needed to provide differing levels of functionality such as EDI for automated exchanges with newer Web-centric technologies, which include XML for automated exchanges and Web interfaces for partners.

Improving Internal Processes – Bringing improvement in an enterprise's internal processes happens to be an very important factor for EAI. EAI techniques can be used to simplify information flow between departments and divisions of the enterprise. In some organizations, EAI provides integrated information for decision making process. EAI can be used to populate data warehouses towards analyzing market trends, evaluating the effectiveness of a business initiative, and assessing the performance of organizations within the enterprise. EAI makes easier the construction of a data warehouse by mediating the flow of information from stove-pipe applications to the common warehouse and by supporting the conversion of data from various applications' formats to a common format.

Employee self-service is also a vital application in improving business processes. Web-enabled interfaces may provide employees with better access to the information they need to do their jobs effectively. Employee self-service Web sites for benefits of administration and other Human Resource (HR) functions are getting popular as days go by. Finally, EAI helps to eliminate manual steps in business processes and to avoid redundant entry of data. Such applications of EAI often employ a work-flow-automation tool to bridge between the applications that are being integrated.

Reducing Time to Market – Information Technology (IT) organizations are finding it increasingly hard to maintain developing mission-critical applications on time and on stipulated budget. There are various difficulties in accomplishing these very important requirements. EAI technology comes to the rescue by reducing the time to market new applications [2]. EAI contributes to faster roll-out in several ways. First, EAI leverages the capabilities of existing applications. For example, often the existing code does its job very well and it has been debugged. The only requirement is to make the functionality by this code accessible to new front-end channels like the Web or to new composite applications. EAI helps immensely in this regard and shortens the time to market. Thus, having the EAI architecture in its place, enterprises can reap the benefits in the long term.

Integrating functionality that is hosted on diverse hardware and operating system platforms is really the trickiest and most error-prone task. With the arrival of CORBA, an EAI middleware, this task becomes relatively easy. Again, EAI reduces the amount of code to be written for an application and helps developers to concentrate on the business aspects of the mission-critical application rather on concentrating on infrastructure. There are maintenance benefits as well.

10.1.6 EAI of Legacy and Packaged Applications

An enterprise's legacy applications have become the repositories of the enterprise's corporate knowledge. Business rules embedded in legacy applications are often documented nowhere else. As a result, it can be extremely difficult to replicate or re-engineer the legacy applications. Thus, utilizing these applications to support new requirements or to design newer applications out of these by integration technologies is really a tremendous advantage and challenge also. Also, legacy applications that run on mainframe computers have many additional advantages. Mainframe computers are ultra-reliable and can support extremely high transaction rates and a large number of users simultaneously. Thus keeping applications on mainframes can help spread the high fixed costs of that computing across a large number of applications.

In the recent past, most of the organizations have made very high investment to deal with the Y2K problem in the mainframe computers and that resulted in much better understanding and better documentation of their legacy applications. That is why, legacy applications are there as jewels for the enterprises. Thus EAI is bound to play a vital role in extracting the critical components from these legacies and to design new applications to meet the advanced requirements from the customers.

Packaged applications, such as Enterprise Resource Planning (ERP) packages from SAP, PeopleSoft, Oracle and other vendors, also figure prominently in the EAI landscape. In the last decade, utilizing of packaged applications has grown significantly as they can provide proven solutions to common business needs. Development of maintenance becomes the problem for the vendor only and this helps the IT staff in the enterprises to concentrate on the problems specific to their business. Here comes the need for EAI as packaged products from a single vendor can not meet all the requirements of an enterprise and these products should be integrated with enterprise-specific custom applications and sometimes with other packaged products. Thus nowadays, having realized the importance of integration process, package vendors started to seek the help of EAI technologies to make their products EAI-compliant.

10.1.7 The World Wide Web and EAI

The Web has become a key factor in the emergence of EAI as an important technology. There are several Web-based applications that depend on EAI. The arrival of Web and its related Internet technologies are fundamentally necessary for EAI. Also the Web happens to be a real revolution and has become the largest information base. There came technologies to disseminate, communicate and share the information to anyone at anytime from anywhere. The Web has brought quite a lot of innovations in commerce and trade, information exchange, etc.

From the perspective of EAI, several aspects of the Web revolution are significant. The Web provides universal connectivity, creates a whole new arena for business explosion and ultimately gives the controlling power to the users. Thus, anyone with a Web browser and an Internet connection can view the Web sites of those who are in the field. Also a business can offer its customers revolutionary products and services through the Web and every business is being compelled to join the race to offer those products and services. This provokes competition among the enterprises and results in a number of benefits, such as extraction of the best service and price, for the users, such as employees, supply-chain partners and customers.

The Web affects any enterprise in two ways. One is, a business can offer its customers revolutionary new products and services through the Web and the second one is virtually every business is compelled to join the race to offer those new products and services to face the competitions. The Web also can offer a new and better way to perform an existing service. The Web technology facilitates some novel services that are otherwise impossible. Business partners and consumers can meet and exchange goods and services through electronic markets. Business partners can greatly improve their coordination through Web-based exchange of information, streamlining the flow of goods and paperwork and squeezing costs out of the supply chain.

The Web has opened the floodgates of customer expectations. Customers expect to find a business on the Web. They expect to get information about products and services being offered through the Web. The Web also enhanced customer relationships. Customers like the feeling of control that they get from the best self-service sites. Businesses also like customer self-service because it can significantly reduce the cost of transactions.

Web-based application integration has moved from the experimental phase. It is expected that this process is all set to gain momentum in the days to come. Effective Web-based applications often depend on the integration of existing applications. The information and services that today's enterprises want to exploit in their public Web sites and their corporate intranets are locked up in existing IT applications. Thus EAI is a must to build Web applications that are competitive. Thus the emergence of Web has provided a crucial incentive for enterprises to utilize the benefits of EAI.

10.2 Characteristics of Enterprise Application Bus

10.2.1 The Enterprise Service Bus: Making Service-Oriented Architecture Real

The Enterprise Service Bus (ESB) is the infrastructure which underpins a fully integrated and flexible end-to-end Service-Oriented Architecture (SOA). This chapter details the essential metadata and capabilities of the ESB. It presents a summary of the key concepts of the ESB and defines the integration model for it, including key user roles. These roles are fulfilled using metadata that describes the service endpoints, such as the service interface and policy requirements and capabilities. The ESB manages this metadata through a registry, which supports configuration, connection, matchmaking, and discovery of service endpoints. Some typical mediation patterns that are used to satisfy endpoint policies are explored, and usage patterns are described in which the ESB is used to implement real SOAs.

10.2.2 Essential Characteristics of an Enterprise Service Bus (ESB)

The metadata that describes service requestors and providers, mediations and their operations on the information that flows between requestors and providers, and the discovery, routing, and matchmaking that realize a dynamic and autonomic SOA. The ESB provides the tools and runtime infrastructure to realize the promise of SOA formulated in the iconic 'publish-find-bind' triangle that was popular in the early days of the SOA revival caused by Web Services. The ESB manages and exploits metadata describing interaction endpoints as well as the domain models used to describe the capabilities of those endpoints, it supports configuration of links that bridge between capabilities demanded by service requestors and those offered by service providers, dynamically matching requestors with providers and in the process establishing and enacting contracts between those interaction endpoints.

10.2.3 ESB in a Nutshell

The ESB enables a SOA by providing the connectivity layer between services. The definition of a service is wide and it is not restricted by a protocol, such as SOAP (Simple Object Access Protocol) or HTTP (Hypertext Transfer Protocol), which connects a service requestor to a service provider; nor does it require that the service be described by a specific standard such as WSDL (Web Services Description Language), though all of these standards are major contributors to the capabilities and progress of the ESB/SOA evolution. A service is a software component that is described by metadata, which can be understood by a program. The metadata is published to enable reuse of the service by components that may be remote from it and that need no knowledge of the service implementation beyond its published metadata. Of course, a well designed software program may use metadata to define interfaces between components and may reuse components within the program. The distinguishing feature of a service is that the metadata descriptions are published to enable reuse of the service in loosely coupled systems, frequently interconnected across networks. What do we mean by 'publishing' a description of a service? Descriptions of the services available from a service provider can be made accessible to developers at the service requestor, possibly through shared development tools. The ESB formalizes this publication by providing a registry of the services that are available for invocation and the service requestors that will connect to them [3].

The registry is accessible both during development and at runtime. Components such as J2EE, EJBs or database-embedded functions may be published as services, but not every J2EE, EJB is a service, and not every J2EE EJB is accessible by means of the ESB. In general, EJBs need additional metadata, and possibly additional bindings, published to the ESB registry in order to make them available as services. Publication of the service requestors and providers allows their metadata to be administered through the ESB registry and enables their relationships and interactions to be visualized and updated.

Nonetheless, ad hoc requestors and providers may also connect to the ESB without first being registered, for example, subscribers to a 'publish/subscribe' topic. In that case, their interactions will not benefit from the full dynamic capabilities of the ESB, described later.

The ESB populates the registry with metadata about services in three different ways. When services are deployed to the runtime environment, they can be simultaneously and dynamically added to the ESB; metadata associated with components already deployed can be explicitly added to the ESB; or the ESB can discover services and service interactions that are already deployed and incorporate metadata describing them in the registry. Note that the ESB is the infrastructure for interconnecting services, but the term ESB does not include the business logic of the service providers themselves nor the requestor applications, nor does it include the containers that host the services. Hosting containers and free-standing applications are enabled for interaction with ESBs with varying levels of integration, depending on the range of protocols and interoperability standards supported. Most containers (e.g., J2EE application servers, CICS*, Microsoft .NET**) integrate with an ESB across the SOAP/HTTP protocols, but fewer have direct support for SOAP/JMS (Java Messaging Service) over a particular brand of JMS provider. After the ESB has delivered its payload to a container, its responsibilities are fulfilled. Within the container, the service invocation may be redirected among the machines in a cluster, or it may be responded to from a local cache. These are some of the normal optimizations within an application

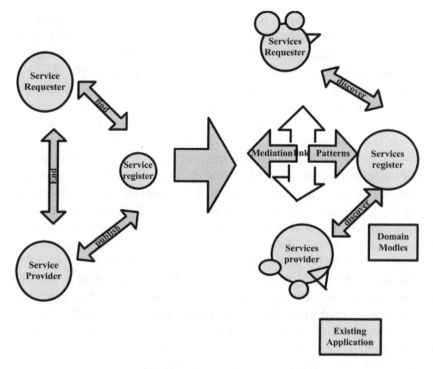

Figure 10.1 ESB underpinnings for SOA.

server environment, and they complement the routing and response capabilities of the ESB between the service providers it interconnects. Similarly, the ESB is the connectivity layer for process engines that choreograph the flow of activities between services. The process engine is responsible for ensuring that the correct service capabilities are scheduled in the correct order. It delegates to the ESB the responsibility for delivering the service requests, rerouting them if appropriate. A core tenet of SOA is that service requestors are independent of the services they invoke. As a result, it is not surprising that the ESB is essentially invisible to the service requestors and providers that use it. A developer can use an API (application programming interface), such as JAX-RPC (Java API for XML-based RPC [remote procedure call]) to a Web service, or distribute messages with the Web-Sphere* MQI (Message Queue Interface) to a message queue, without considering whether these requests are flowing directly to the service or traversing an ESB.

Similarly, a service provider can be written as a J2EE EJB or a servlet without any specific application code to make it accessible through an ESB.

Despite this, one of the values of the ESB is that it takes on the responsibility for many of the infrastructure concerns that might otherwise surface in application code. Thus, although developers can use APIs for service invocation, they do not need to add logic to deal with security, for example. The ESB virtualizes the services that are made available through the bus. The service requestor, both in its application logic and in its deployment, does not need to have any awareness of the physical realization of the service provider. The requestor does not need to be concerned about the programming language, runtime environment, hardware platform, network address, or current availability of the service provider's implementation. In the ESB, not even a common communication protocol need be shared. The requestor connects to the bus, which takes responsibility for delivering its requests to a service provider, offering the required function and quality of service. Not surprisingly, the infrastructure of the bus is itself virtualized, allowing it to grow or shrink as required by the network and workload which it is supporting.

The flexibility that comes from a SOA, and the virtualization it implies, is fully realized by the dynamic nature of the ESB. All the metadata, conditions, and constraints used to enable a connection from a requestor to a provider can be discovered, used, and modified at runtime. For example, a new implementation of a service in a different geographical region can be published to the ESB registry, and requests in that region can be routed to it without reconfiguration of the requestors. A service requestor might select a reduced level of assured delivery and see an improved level of performance as the ESB determines that it can use a different delivery protocol. This flexibility is available as a direct consequence of the role of the ESB registry. Because all relevant metadata for the service provider and service requestors has been placed in the ESB registry, it can be subsequently discovered and used to make dynamic changes. To achieve much of this flexibility, the ESB accepts requests as messages, then operates on them, or 'mediates' them, as they flow through the bus. Mediations can be an integral part of the ESB, providing transport mapping between SOAP/HTTP and SOAP/JMS, or routing a message to an alternate provider if response times fall below an acceptable value. It is also a feature of the flexibility of the ESB that mediations can be provided by third parties – by other products, ISVs (independent software vendors), or customers – to operate on messages as they flow through the ESB infrastructure. This allows, for example, ISV packages to implement advanced load-balancing features among services, or customers to add auditing to meet new legislation. Mediations can be deployed on the ESB without changing the service requestor or provider.

Mediations are the means by which the ESB can ensure that a service requestor can connect successfully to a service provider. If a service provider requires one format for an address field and a service requestor uses a different one, mediation can map from one format to another so that the ESB can deliver the service request. If the service provider expects encrypted messages, a mediation can encrypt the 'in-the-clear' service requests as they pass through the ESB. The ESB can react dynamically to the requirements of requestors and providers when they are described in their metadata and held in the registry. In the case of message formats, this is usually achieved through a schema definition. For other service properties, policy statements, which may describe the encryption algorithms to be used or the requirements for auditing, can be associated with the metadata of the service provider and requestor. The ESB consults this metadata at runtime and can reconfigure the mediations between requestor and provider to match the requirements. By annotating a policy for the service providers in the ESB registry, the system administrator can, for example, ensure that the services meet the company's new privacy guidelines. Thus the ESB implements an autonomic SOA, reacting to changes in the services it connects.

One of the major uses of mediations is in systems management. Mediations can be deployed in the ESB environment to enable request and response messages to be monitored as they flow through the system, enabling service-level management or problem determination. Mediations can route service invocations to back-up data centers if there is a local problem or to new service providers as they are brought online. They can validate messages in terms of their format correctness, data values, or user authentication and authorization. Through these and other systems management capabilities, the ESB ensures that a loosely coupled and dynamically varying SOA is still manageable in a production environment.

Many of the mediation capabilities just described are core attributes of the ESB, and the mediations are made available as part of the runtime environment. They are customizable, so that, for example, a generic table-driven routing mediation can be configured to use a specific table and a specific field in a message as the key. The ESB also provides tools to configure the interactions between services – to display the services available in the ESB, to interconnect them, to add policy requirements to a service or group of services, to identify mismatches in the endpoints, and to associate mediations to correct these, either explicitly or through automatic reconciliation of their policy declarations. Much of the preceding discussion uses the terms service requestor and service provider, as is appropriate for the ESB. Service requestors and service providers are equal partners in the interaction, with the requestor simply being the endpoint that initiated the interaction. The interaction may continue with either endpoint sending or receiving messages.

The ESB supports many different types of program interaction: one-way messages as well as requests and responses, asynchronous as well as synchronous invocation, the publish/subscribe model, where multiple responses may be generated for one subscribe request, and complex event processing, where a series of events may be observed or consumed to produce one consequential event. The ESB is also, in principle, transport and protocol 'agnostic', with the capability to transform messages to match the requestor's preferred formats to those of the provider. In practice, most ESBs support SOAP/HTTP, which reinforces its role as an interoperability standard. They also support a range of other transports and protocols, some for use by service requestors and providers connected by the ESB, and some for internal communication within the ESB.

10.2.4 The ESB Integration Model

Having established the basic concepts and features supported by the ESB, we focus on ESB-based SOA solutions. The ESB integration model captures those aspects of the overall solution that are relevant for the ESB tools and infrastructure to facilitate interactions between ESB-managed service endpoints. It enables the various user roles involved in creating and managing those solutions to express the ESB relevant information that they contribute or monitor, and the ESB runtimes to manage interactions accordingly. In essence, the model contains metadata describing service endpoint requirements, capabilities, and relationships, including information describing the specific details of interaction contracts. Not all user roles use the model directly. A business analyst, for example, might define a set of key performance indicators (KPIs) that need to be translated into events produced by the underlying implementation artifacts and potentially into parameters of a service interaction contract. Users in the architecture and design space, such as a solution architect, might define service capabilities and requirements in more abstract terms than those required by the ESB integration model. Other roles, such as the application developer, create and use service capabilities and requirements for the integration model. The user roles that are the most interesting from the ESB perspective are those of the integrator and the solution administrator. We next discuss how integration specialists assemble solutions from existing service components and how solution administrators configure and reconfigure those solutions. We use UML (Unified Modeling Language) models to illustrate the concepts those user roles deal with and these are high-level conceptual models not to be confused with product- or implementation-specific models of the underlying runtimes or specific standards in this space. We do, however, hint at relationships to products and standards where appropriate. Integration specialists assemble business solutions from a set of service components. They do not have to understand the implementation details of those components (process coordination, existing applications, interactive tasks, etc.), all they need to understand is the capabilities offered by components and the requirements of the components they use, with respect to other components.

The ESB service registry provides the required information about those components and, together with the ESB runtime, enables integration specialists to perform component-assembly tasks, selecting components required to implement the solution, resolving dependencies those components might have on other components, and interposing the mediations required to make components interact. Solution administrators deploy and customize the solutions they get from their integration specialist colleagues: they may be given a set of component relationships which they simply adopt; they may choose to override the defaults defined by an integration specialist; they may have to compensate for the fact that an integration specialist has not resolved certain variables of a solution; or they may have to reconfigure a previously deployed solution due to changes in the solution environment. These tasks are usually done in the component development environment, but the ESB enables flexible configuration through late binding by providing this service to the solution-administrator role. The key to enabling this flexible configuration and reconfiguration of solutions is the explicit declaration of capabilities and requirements of service interaction endpoints. The next section explores the underlying service metadata management capabilities provided by the ESB in more detail.

10.2.5 SOA Metadata

The key to service virtualization and dynamic matchmaking between service requestors and providers is the explicit declaration of capabilities and requirements of interaction endpoints. Service metadata describes capabilities of software assets independent of their implementation specifics. It does not assume a specific programming model for the realization of the services that offer their capabilities for use by other components, nor for the realization of services that require certain capabilities to be provided by other components. It facilitates interoperability among a broad spectrum of service providers and requestors; an existing CICS/COBOL application can declare its service capabilities and expectations exactly like a business service newly implemented in J2EE or a Web service offered by a business partner that uses SOAP/HTTP. Service metadata also supports an 'assembly-from-parts' model for implementing new business process applications from a catalog of services. Before discussing the kind of metadata about a service that is relevant for ESB-managed interactions in detail, it is important to note that requirements for explicitly recording metadata about service interaction endpoints are relative to the scope of visibility of the underlying application artifacts (Figure 10.2).

Even in moderately homogenous development environments (e.g., departmental integration projects using messaging features of an application server) it can make sense to use the service component abstraction and an ESB to construct solutions. In that case, a minimal set of metadata declarations (e.g., interface declarations only) for each service component is sufficient to enable ESB-facilitated service component assembly. If the principles of SOA are applied on a larger scale (on an enterprise level, using a message broker), more explicit declaration of service component capabilities and requirements are necessary. In this case, not only service interfaces but also quality-of-service assumptions might have to be considered to enable service users to understand under which circumstances they can use a service. An even more detailed declaration of service capabilities is required when attempting to apply SOA on an inter-enterprise scale. In this case, declaration of interaction patterns, service-level agreements, and other factors become relevant. Figure 10.2 also illustrates the fractal (self-similar) nature of the ESB. Various ESB agents collaborate to realize an integration infrastructure that enables a business process created in a departmental context to interact with service that may be local to the department, hosted in other departments of the same enterprise, or even hosted by other enterprises. Having established the relativity of metadata needs for various levels of ESB elements, we take a more systematic look at the spectrum of metadata that could be recorded for a service component. As established earlier, the main elements of the ESB programming model are service requestors and service providers. Most of the ESB-relevant metadata about those interaction endpoints can be classified as interface declaration or policy annotation. The objective in recording metadata is the establishment of matches between potential partners so that they can interact. A given requestor matches a particular provider if their interface and policy declarations are compatible or can be made compatible.

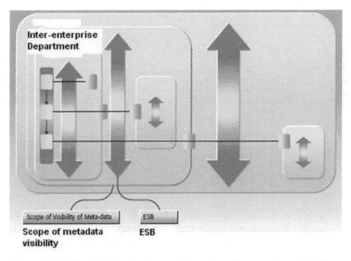

Figure 10.2 Scope of service assembly and service metadata visibility.

The simplest form of compatibility is a perfect match. In this ideal case, a requestor needs the exact operation a provider offers with the qualities of interaction service defined by the provider's policies. The only practical way to make this work is the 'requestor makes it right' principle that is common practice in traditional RPC-style interactions. Using this principle, the application developer of a requesting application consults the ESB service registry to find a provider and implements the requestor to fit the provider's specification. In SOA scenarios, a more flexible matchmaking model is required. The ESB supports such a model, but this comes at a price; namely, potential interaction partners need to provide sufficient information about their requirements and capabilities. The simplest and most common form of service metadata is the interface declaration. Interaction endpoints describe the messages they can process or will produce, as well as the message exchange patterns supported. In Figure 10.3, service providers support and service requestors require support for service interfaces that are made up from service operations, which in turn represent message interchanges in terms of logical messages supported by the endpoint. Ideally, the interface part of a declaration of service capabilities and requirements is described by using WSDL, with XML schema describing the structure of messages to be exchanged. Nevertheless, there are many examples of successful ESB implementations that use only a subset of WSDL (e.g., XSDs for message declarations for processing information, often with specific annotations to capture additional

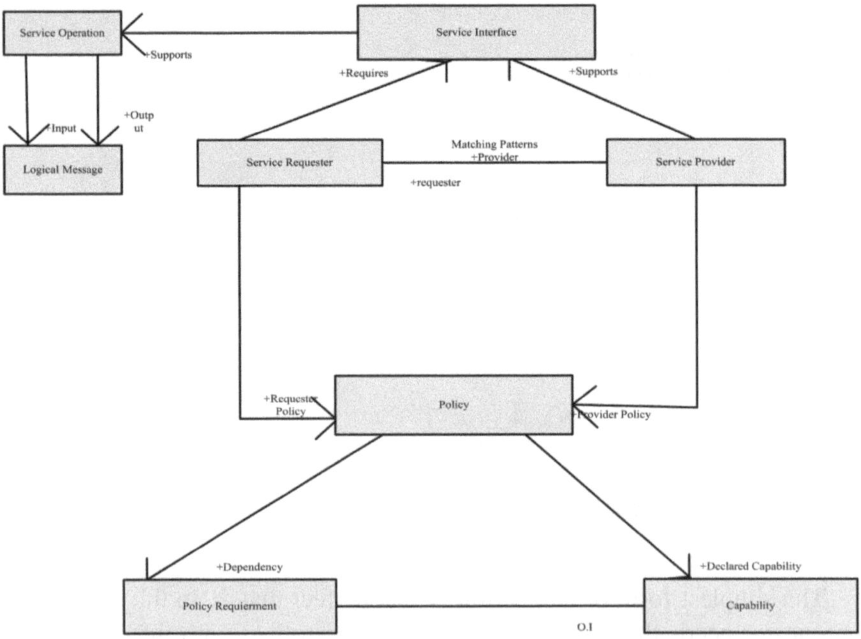

Figure 10.3 Service capability and requirements declaration.

metadata about message formats) or home-grown metadata schema to capture the information. Those approaches, in general, do not scale beyond relatively homogenous, 'localized' ESBs, but they support the point that it is more important to make the capabilities and requirements of services explicit rather than use a specific declaration formalism. Policy declarations further qualify capabilities of interaction endpoints; simply put, a policy expresses anything a component wants the world to know about it other than what messages it understands. This is a very general concept of policy, and it actually covers things that are sometimes factored out into behavior declaration or semantic annotation of services; as explained previously, we present a very coarse-grained, conceptual model here that may well be refined during translation into an operational model. As discussed previously, in many localized ESB scenarios, policy declarations are not required.

In the general case, interaction endpoints use policies to declare the capabilities they offer or policy requirements they have with respect to other services. Note that the model illustrated in Figure 10.3 allows both service providers and requestors to declare capabilities and requirements by use of a policy. A requestor can declare not only requirements for providers that wish to interact with it, but also capabilities which specify actions that it is willing to perform before it dispatches a request to a provider (e.g., encrypting messages). Service providers use policies to declare not only the capabilities they offer, but also the requirements they might have for requestors that want to use them. More precisely, policy declarations can provide (1) information about characteristics of the declaring component's internal behavior which affect the interaction, (2) constraints on a peer's invocation of the declaring service, (3) constraints on the target interface of an interaction partner, or (4) information about characteristics of the service component's internal behavior regarding the respective interface reference. The WS-Policy declaration establishes a framework for policy declarations of all sorts and enables reasoning about the compatibility of policies declared by potential interaction partners.

The policy declarations discussed here should not be confused with policy declarations intended for the service container that hosts the service component in question. As indicated earlier, we use a very broad definition of policy and include declarations a service requestor or provider might want to make about expected or supported behaviors. For example, a provider might indicate sequencing constraints on its operations describe interaction patterns involving more than one operation or maybe even describe more than one interaction partner. Application developers can represent such behavioral specifications, for example, using the features for abstract process declaration provided by (Business Process Execution Language (BPEL). Note that they will in general not expose the actual behavior of their services but rather a projection or abstraction of that behavior.

In the very broad definition of policy here, we also include semantic annotations of service interfaces that explain the meaning of messages exchanged with the service or the meaning of the service operations. The annotations are useful when the semantics of services and their operations is not obvious from the naming conventions used. The annotations can be as simple as relating elements of the service declaration to well-known terms, but they can also include declarations of preconditions or post conditions for operations.

We describe a conceptual model for the ESB, not a particular implementation. Service interfaces can be captured in a variety of ways, and from an abstract ESB perspective, the specific syntax used to describe them is far less important than the fact that they are recorded. Declaration of message sets using XML Schema, with annotations to capture information relevant to a specific message formatting, is an example of ESB-managed metadata about interaction endpoints (in this case WebSphere MQ applications that produce and consume those messages). Where possible, however, we encourage use of the WS (Web Services) standards to declare capabilities and requirements of interaction endpoints: WSDL for service interface declarations, WS-Policy for any kind of policy annotation to those interfaces, and BPEL for specification of sequencing constraints. Standards play an important role in advancing the syntactic normalization of metadata about interaction endpoints. Nevertheless, in many cases, especially in ESB scenarios beyond departmental applications, deeper understanding of the semantics of the underlying services is required to perform any meaningful matchmaking. Service interfaces abstracted from a legacy application might make sense to requestors that know about the original application, but not to requestors outside of that small circle; often even a 'perfect match' between services on an interface level does not imply that they can actually interact in any meaningful way.

As indicated earlier, the semantics of services is often declared through reference to more commonly known concepts, captured in domain models. Domain models establish a frame of reference that makes it possible not only to explain the semantics of services but also to make statements about compatibility of services, and that in turn enables the definition of service interaction contracts which can be managed by the ESB infrastructure.

10.2.6 The ESB Service Registry

The ESB plays two main roles in the service endpoint matchmaking game – the service registry manages all relevant metadata about interaction endpoints (Figure 10.4), and it also takes care of the matchmaking between those endpoints. This section discusses the service registry; the next section discusses the matchmaking. The ESB is a service registry in the sense that it manages not only metadata about the service interaction endpoints involved in the SOA, but also information about domain models. This information establishes a common understanding of services beyond the scope of visibility of an individual service requestor or provider. The ESB captures information that can be used to better understand the practical content of the registry: domain models representing general knowledge about a topic area, independent of the specific domain applications represented as services in the registry. As before, we use a very generic definition of the term domain model because we want to establish an implementation-independent model for the ESB. Our definition covers domain models as simple as a topic space or a simple taxonomy that classifies events exchanged in publish/subscribe style interactions, it includes standard message sets used in specific industries or a set of 'generic business objects' covering a specific application domain; and it extends to moderately complex ontologies describing concepts and their relations in a particular topic space. In the ESB integration model, domain models are used to establish semantics of the practical metadata artifacts that the ESB cares about.

In many 'local ESB' scenarios, little or no domain knowledge needs to be formalized – the user community involved simply knows the semantics, and a simple hint in the form of well-named interfaces and messages suffices. In the publish/subscribe messaging model, topic spaces can be used to classify message instances. Generic business objects or standard message sets applicable to a domain can be used to establish a (semantically) normalized view for messages exchanged within that domain (but usually not beyond it), thus enabling application developers to implement endpoint applications without worrying about the specific semantics of the underlying service components. This enables service-level managers to observe the status of a system, based on knowledge about events which adhere to the Common Base Events (CBE) standards proposal. Ontologies can establish a deeper semantic understanding of the interaction endpoints and facilitate use of more sophisticated ways to identify possible interactions between them. The main objective of managing domain models and establishing semantic understanding of interaction endpoints is to enable matchmaking between those endpoints – the more explicit knowledge there is about capabilities and requirements of the endpoints, the more automated the matchmaking can be. The main role of the ESB registry is to manage metadata about the interaction endpoints themselves. To participate in ESB-managed interactions, endpoints need to register with the ESB. The ESB model represents registered service requestors as bus service requestors (BSRs) and registered service providers as bus service providers (BSPs). Service providers that are not registered as BSPs are invisible to the ESB for interaction partner selection.

When a BSP or BSR is registered, its service interface and policy declarations are captured in the ESB service registry. At this point it is also possible to provide additional information about the service endpoint that might not have been provided with the original declaration. Semantic annotation is one example. Documenting or discovering relationships between the newly registered service and other artifacts in the service registry is another. Both BSPs and BSRs can be created without any available 'counterpart' that wishes to interact with them; conceptually, the ESB ensures that it will take

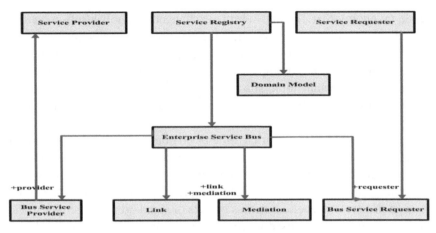

Figure 10.4 ESB service registry content.

care of connecting each BSP to requestors that send requests to it and that it will deliver the requests of each BSR to some matching provider. In a way the ESB is the 'ideal provider' for a BSR and the 'ideal requestor' for a BSP. It is the ESB's job to make things right if there actually should not be such an 'ideal counterpart'. The ESB registry also holds details of links and mediations, which are described in the next subsections. Like any good service registry, the ESB provides the following features:

1. Discovery and management of meta-information about interfaces and capabilities of existing applications that can be used as building blocks for integration solutions. This includes analysis of legacy applications to discover meta-information about their interfaces, policies, and behavioral constraints, as well as exploitation of object discovery agents to capture meta-information about packaged applications.

2. Management of meta-information about services. This includes WSDL declarations as well as WSPolicy declarations describing capabilities provided by services or required by service requestors and also BPEL-defined declarations of behavioral constraints for services (abstract BPEL processes) or actual behavior of those services.

3. Management of domain models describing general knowledge about an application domain relevant for SOA-based business integration scenarios. Examples include industry-standard message sets, generic business objects, ontologies encoded in the OWL Web Ontology Language, and 'contracts' for SOA interactions.

4. Discovery and management of relationships between 'real world' artifacts representing existing applications, service declarations, and domain models. Enabling generation of artifacts for the ESB runtime. Examples include Service Data Object (SDO) schema, maps between service interfaces for ESB mediations, and application adapters.
5. Management of runtime meta-information for matchmaking between service requestors and service providers (e.g., service declarations with policies and compatibility rules) and SDO schema with annotations.

10.2.7 Links and Mediations for Dynamic SOA

The ESB supports two concepts to facilitate interactions between endpoints: it introduces links between service requestors and providers that enable basic connectivity between interaction endpoints with a configurable quality of service, and it provides the concept of mediations that can be used to configure and reconfigure the ESB by dynamic alterations to routing and qualities of service and to allow interaction endpoints to modify their behaviors. Links and mediations basically realize the contract between interaction partners that is implicit in the declarations of the capabilities and requirements of those partners. An ESB link has two endpoints, one for attachment of BSRs and one for attachment of BSPs; both ends can be qualified by using interface declarations and policies just like interaction endpoint declarations. A link defines the 'ideal counterpart' for service requestors and providers: the provider attachment part of a link can be designed to provide an exact match for a particular registered BSP; and the requestor attachment part of the same link can be designed to provide an exact match for some registered BSR. The two ends of the link do not have to match – ESB-managed mediations on the link can enact the transformation required to make the ends meet. Thus, an ESB link represents and implements the contract between interaction endpoints. It can be tailor-made for a particular requestor/provider pair. An integrator given the task to resolve the requirements of a business process component (here in the role of a service requestor) for services that implement process activities might select a set of service providers to be linked to the process and, in cooperation with the solution deployer, create a BSR representing the process, a set of BSPs representing the services invoked, and a set of links between them, configured to meet the requirements declared by the process. Link configuration does not necessarily reflect only the requirements and capabilities of the endpoints that it connects; it can just as well implement requirements defined for a set of interactions, for example, in enterprise policies (such as logging all high-value transactions). A solution administrator can pre-configure ESB-managed links to support a specific quality of (interaction) service to be used by a number of interested interaction endpoints. A link might be configured with a varying number of interaction endpoints attached to it. A solution administrator might register a BSP and attach it to a link that might not have any BSRs attached to it. Alternatively, the administrator might register an event source as a requestor and attach it to a link that will propagate the events it produces – potentially, without anybody listening at the other end. A link can be configured such that it dynamically determines the endpoints that need to be attached to it, e.g., depending on the content of the requests it is processing [4]. In the most dynamic case, a link can be created dynamically to perform matchmaking between dynamically established requestor/provider pairs.

The configuration of the link can be derived from the requirements and capabilities of the endpoints that it is meant to connect. As illustrated in Figure 10.5, both endpoints of a link as well as the link itself can carry policy declarations.

Policies on the endpoints declare the constraints as well as the capabilities of the endpoints, whereas policies on the link itself can only represent capabilities implemented by the link for the benefit of attached endpoints that require the capability. Interfaces on the endpoints of a link are optional and often are omitted to indicate that the link can carry any type of message between associated endpoints. Note that service interfaces on both ends of a link can have different signatures. In this case, some mediation on the link needs to take care of the transformation.

SOA holds out the promise that services can be discovered from service directories and bound together to form new and exciting, or simply more efficient, applications. Unfortunately, existing applications were seldom designed to be linked together, and the integration specialist is faced with mismatches of protocol, format, and quality of service to reconcile, as well as requirements to make his new applications more flexible and resilient. The ESB addresses these challenges by interposing mediations between service requestors and providers which can reconcile their differences. In addition, mediations can reconfigure the links between requestors and providers, for example, to create an alternative routing or to create a reactive and autonomic system.

The ESB integration model supports the attachment of mediations at various points on a link between interaction endpoints: mediation can be associated with the registered service provider, registered with a service re-

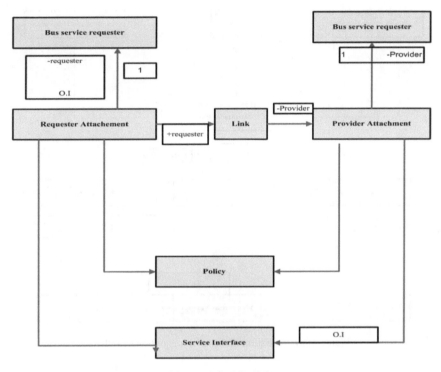

Figure 10.5 ESB links.

questor, or attached to a link between them. This is formalized in the model through the concept of a mediation point (Figure 10.6). A mediation point inserted at the requestor implies that the mediation will be performed no matter what provider the requestor interacts with; a mediation point activated at the provider end implies that the mediation will be performed whenever the provider receives a request, no matter which requestor it comes from; and a mediation associated with a link applies only to the specific interactions that occur through the link. Mediations process messages as they flow through the ESB. Interface mediations operate on the message payload, which contains the information required by the service provider, and can change its content and its structure. In addition, the messages have contextual information associated with them, usually specified in message headers. This concept is familiar from SOAP headers, which contain, for example, the location of the service provider. Within an ESB, the message context includes additional quality-of-service and routing information about the link and the mediations

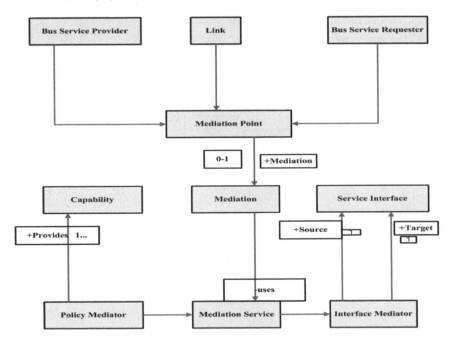

Figure 10.6 Mediation in the ESB integration model.

required between the service requestor and provider. Policy mediations operate on the message context. In addition to the information in the messages, the ESB provides mediations with information about the ESB configuration by means of access to the ESB registry.

Mediations can change the content and format of the service requests or responses or modify their intended routing. Although these are useful and necessary functions of mediations, to fulfill the flexibility of an ESB, it is also important that mediation can be restricted, for example, so that it cannot see sensitive payload information or change a particular routing. Mediations are characterized by the read or update access they require to the context and payload sections of the message, and this can be enforced by the ESB.

10.2.8 Mediation Patterns

Mediations are not formally restricted in their capabilities, but their intended role is satisfying integration and operational requirements within the infrastructure, rather than implementing business level processes. Within this role, there are several basic patterns, which are seen repeatedly, either as individual mediations or within more complex mediations. The monitor pattern is used to observe messages as they pass through the ESB without updating them in any way. Mediations that conform to this pattern use the information in the context and payload in many different ways; for example, to monitor service levels, to assist in problem determination, to meter usage for subsequent billing to users, or to record business-level events, such as purchases above a certain dollar value. This pattern can also be used to log messages for audits or for subsequent data mining. The first two examples would require read access to only the context information in the message. The others would require read access to the payload too, though the messages could be logged as raw byte streams, for later parsing.

The transcoder pattern changes the format of the message payload without changing its logical content. For example, it may convert a SOAP message into a JMS/Text message with an XML payload that matches the body of the SOAP message (possibly by mapping SOAP header fields into JMS Properties). Mediations which conform to this pattern can often be created automatically when there is a clear definition of the two formats and of the relationship between them. This pattern requires update access to the payload. The modifier pattern updates the payload of the message without any change to the context information. It requires update access to the payload. There are two common sub patterns: transformation and enrichment. In the former, the message payload is transformed from one format (schema) to another, to match the definition of a message of the requestor to that of the provider. This includes 'enveloping and de-enveloping' (the process of putting a message in one network format inside the format envelope needed for transmission over another network, or the corresponding removal of an envelope) and encryption. In the latter, the payload of the message is updated by adding information from external data sources, such as customization parameters of the mediation, or from database queries.

The validator pattern determines whether a message should be delivered to its intended destination. If not, it may silently ignore the message or may return a rejection response to the requestor. The check can be against the metadata of the message, such as the schema, or permitted values for specific fields. Alternatively, the check can be against side information associated with the mediation, which may relate to one or more fields in the payload of the message or to information held in the message context, such as the origin of the message. This variant of the pattern includes authentication and authorization checks. Depending on the checks involved, this pattern requires read access to the message context or the payload, or both. The cache pattern returns a valid response to the requestor without necessarily passing the request to a service provider. It maintains a cache of requests and their associated responses, and if it recognizes the request, it returns the response directly to the requestor. If the response is not available in the cache, the message is sent to the provider, and the cache is updated with the new response on its return. This pattern requires read access to the payload of the request and response messages and is unusual in that it only applies to request/response interactions. The router pattern changes the intended route of a message, selecting between the service providers associated with the mediation. Simple selection would include routing between two versions of a service, with the percentage routed to the new version being increased by the system administrator as confidence in its capabilities increases. Another example is routing to a local version of a service until it becomes overloaded, then routing to a more expensive remote service. This latter case could also take the importance of the message into consideration, as indicated by a 'gold' user status or the size of a purchase, for example. This pattern requires update access to the context of the message and read access to the payload if it is needed for the routing selection criteria.

The discovery pattern queries the ESB registry to discover the set of service providers that match the requirements of the requestor, selects one of them, and routes the message to it. This is an enhancement of the routing pattern; in this case, the set of possible service providers are not preconfigured at the mediation. Suitable providers match the requestor's message format, the quality of service required, or the protocol supported from the mediation to the possible providers. This pattern allows for more flexible routing, for example in the failover situation mentioned previously, where a new remote data center can be brought online and its services registered, without having to update the configuration of every routing mediation. This pattern requires update access to the context of the message and read access to the payload if it is needed for the routing-selection criteria.

The clone pattern makes a copy of a message and modifies its route. The new message will then have a separate existence within the ESB. This pattern is useful in association with the monitor pattern, where the monitoring logic must not be allowed to delay the delivery of the message to its intended destination; for example, when the message is logged to a database. This mediation requires read access for the payload and update access for the context. The aggregator pattern monitors messages from one or more sources over a time period and generates a new message or event, based on the input it considers. It defines a set of event types in which it is interested and uses aggregation rules to derive a new event. It may simply aggregate a specific set of events, or it may look for patterns in event streams and generate a complex event when a pattern is detected. This pattern is useful in complex event processing scenarios.

Mediations can be explicitly configured by the integration specialist or the solution administrator. The former might apply in the case of modifier mediation, and it would transform from the format of the requestor to that of the provider and vice versa. The latter might apply if the solution administrator wanted to add monitor mediation to a particular link in the ESB to measure performance. The ESB can also configure required mediations dynamically to match the policy requirements and capabilities of the requestor and provider. If a service provider requires encrypted messages, the ESB can configure an encryption mediation for the requestor. If the provider changes its algorithm, the next service request will fail; the ESB will query the provider's metadata, reconfigure the encryption mediation, and reissue the request.

10.2.9 ESB Usage Patterns

The mediation patterns described provide some basic building blocks for the ESB. Higher-level patterns provide a means for describing and defining interactions and component topologies at the system or solution level and help us to see how and where the abstract concepts that we have been describing can be applied to specific implementation scenarios. Patterns enable and facilitate the implementation of successful solutions through the reuse of components and solution elements from proven successful experiences. IBM's patterns for e-business provide one such example and, with specific relevance to the ESB, introduce a set of collaboration patterns that design or describe broad organizational relationships among applications and a set of interaction patterns that describe required behavior in greater detail. The fundamental concept in this case is that of the broker application pattern, in which distribution rules are separated from applications, enabling great flexibility in the distribution of requests and events and reducing the proliferation of point-to-point connections, thereby simplifying the management of the network and system. This basic pattern appears in several variations, and we will briefly consider each of these variations in this section.

Service and event-routing pattern: A request or event is distributed to at most one of multiple target providers (Figure 10.7). Examples may include simple service selection based on context or the content of the request or on more complex models, in which service requests can be routed to particular systems based on availability, workload, or detection of error situations. Service selection may involve the lookup of appropriate service providers in a service registry.

Protocol switch pattern: A routing pattern in which requestors and providers use differing network protocols (Figure 10.8). Examples may include simple mapping of SOAP/HTTP requests onto a more reliable SOAP/JMS infrastructure or mapping between JMS and non-JMS applications. Proxy or gateway pattern: A variant of the routing pattern (or protocol switch) which maps service interfaces or endpoints, possibly providing security functions (authorization and access control) and logging or auditing capabilities (Figure 10.9).

The proxy may also support disaggregation (and subsequent reaggregation) of a single request into multiple component subrequests. Examples of this pattern include service portals, in which a single point of contact is provided for multiple services and the details of 'internal' services may be hidden from the service requestors. Event distribution pattern: Events may be distributed to more than one target provider, based on a list of interested parties that is managed by the ESB. Services that wish to be notified of such events may be able to add themselves to the interested-parties list. An example of this pattern would be the distribution of business events based on CBE through the common event infrastructure.

Service transformation pattern: Requestors and providers use different service interfaces, and the ESB provides the necessary translation (Figure 10.11). This pattern exposes new service interfaces without requiring change or modification to an existing application or service. It may also be used when multiple providers support the same business function but provide different interfaces, allowing this difference to be hidden from the service requestor.

Matchmaking pattern: Another variant of the service routing pattern in which suitable target services are discovered dynamically based on a set of policy definitions (Figure 10.12). This pattern is used in very dynamic environments where there are many hundreds or even thousands of services attached to the ESB, and service implementations may or may not be available when any given request is issued.

These basic interaction patterns may also be used in conjunction with process-oriented interaction patterns. A process or workflow definition (defined by using BPEL or some equivalent language) extends the broker interaction pattern by orchestrating the execution sequence for a number of service inter-actions.

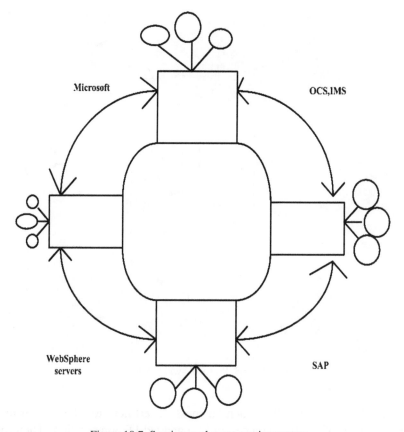

Figure 10.7 Service- and event-routing pattern.

By using these two patterns together, the service that orchestrates the interaction pattern can focus exclusively on business-process requirements, delegating issues of matchmaking, routing, and service selection to the ESB infrastructure.

10.3 SOA and Web Services Technologies for EA Migration

10.3.1 Developing a SOA and Web Service Technology for EA Migration

A lot of large enterprises providing multiple services find today that their IT has been heavily departmentalized. This primarily happened because a lot

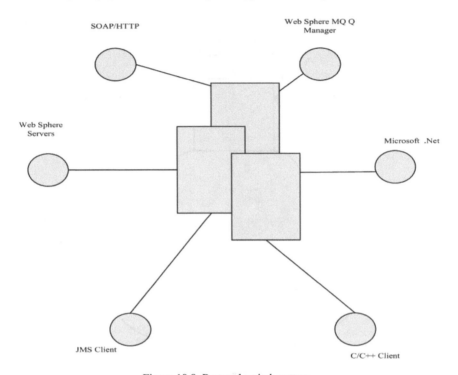

Figure 10.8 Protocol switch pattern.

of enterprises opted to allow each business department to take care of its IT needs, instead of relying on a centrally managed IT organization. Therefore, a lot of departments ended up creating applications in isolation from other departments. Often, a company may have several billing, account managements, and alike systems supporting different aspects of the business. Moreover, a lot of companies tend to install the 'best of breed' software without considering integration with other applications. They later find that special arrangements are often required for making a new application a part of the IT infrastructure. Figure 10.12 presents just an example of an IT infrastructure.

One of the IT challenges that could be noticed from the diagram above is that most of the applications communicate directly to each other. Such dependency may become a real problem when an application needs to be modified or phased out. Any modification may lead to updating each unique communication line in its own way. Therefore, such changes may become a costly endeavor. This situation is called *tight coupling* between applications and is becoming a real head-ache for some enterprises.

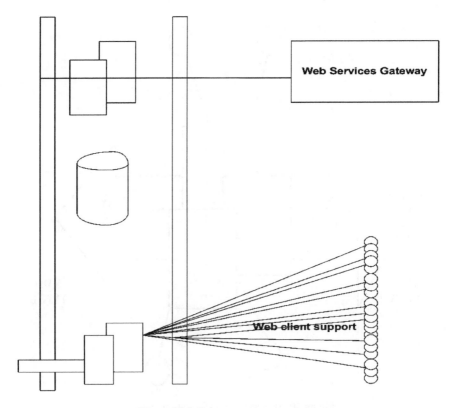

Figure 10.9 Proxy or gateway pattern.

On the other hand, SOA & Web Services put loose coupling as one of the main principles for successful application integration on the enterprise level. Opposite to tight coupling, loose coupling is: restricting the number of things that the requester application code and the provider application code know about each other. If a change is made to any aspect of a service that is coupled, then either the requester or the provider application code (or, more likely, both) will have to change. If a change is made by any party (the requester, provider, or mediating infrastructure) to any aspect of a service that is decoupled, then there should be no need to make subsequent changes in the other parties [5].

Figure 10.12 also demonstrates a practice employed by some enterprises, when an application with the required functionality is deployed in several business areas. Often, the application is slightly modified with each deployment to meet unique requirements of a specific business area. While there

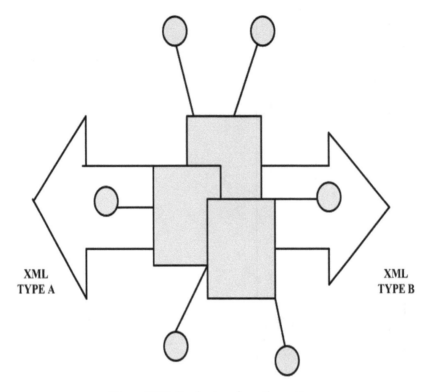

Figure 10.10 Service transformation pattern.

are no obvious disadvantages to having multiple applications with the same functionality, their existence may indicate the following:

- Potential existence of data duplication in the enterprise, which compromises operational data accuracy. This is caused by most of such applications relying on the same data sources and storing portions of that data locally for performance or other reasons.
- Higher cost of maintaining multiple applications than the cost of supporting a single solution.
- Special care needed when such applications are consolidated to reduce impact on applications that have been interdependent with the phased out solutions.

These challenges can be addressed by applying SOA & Web Services principles for a successful integration of applications on the enterprise level.

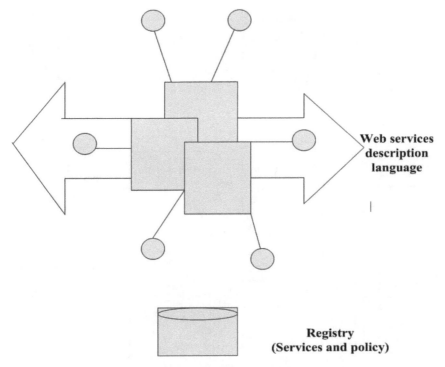

Figure 10.11 Match making pattern.

It is worth noting that when an enterprise is not heavily departmentalized, applying SOA & Web Services may be a reasonable task. However, when a number of IT departments within a company are big and the number of applications that they host is even bigger, implementing the SOA & Web Services-based Enterprise Architecture may become a challenging task that requires careful planning.

Splitting a complex and heavily departmentalized enterprise into a number of separate domains may make things much easier. First, each domain may be decoupled from the rest of the enterprise by exposing specific application services through the ESB, and, second, applications within a domain can be later SOA & Web Services-enabled without affecting the rest of the enterprise.

Such partitioning into domains may be organized according to a business function (for example, Sales, Account Management, Customer Services, and so on) or implementation (such as Mainframes and Unix servers). However,

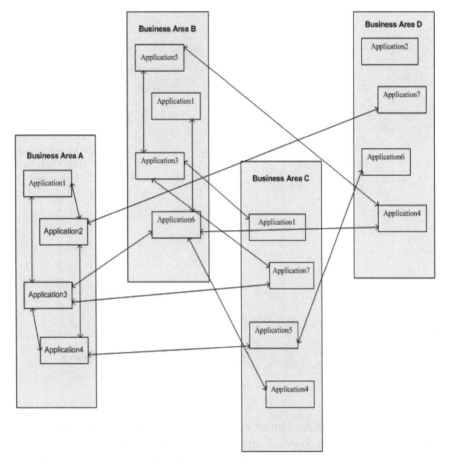

Figure 10.12 Common legacy enterprises IT infrastructure.

essentially, a domain would contain applications that are connected tighter, and would therefore have greater dependencies to each other then to the rest of the enterprise.

The following criteria may be used for determining if a set of services belong to a separate domain:

- *Functional domains* are selected based on a business function. Services in these domains consume a limited number of services from the outside of the domain and expose a limited number of services to the outer enterprise.

- *Technology-based domains* are selected based on a set of technologies they utilize. These may include mainframe applications, distributed applications, and so on.
- *Application-based domains* are tightly coupled collections of applications. e-Business applications or a collection of mainframe applications that share the same database may be an example of such separation.

Initial separation into domains may also be purely conceptual, allowing for identification of application functions that need to be exposed as services to the rest of the enterprise.

Once services are identified, domains need to be separated from each other by establishing clear boundaries enforced through use of gateways and firewalls. Such separation allows for better control over application interactions and further flexibility for making changes to applications without dramatically affecting the rest of the enterprise. Such separation may be achieved through several different approaches:

- Defining *business functions* that expose functionality of a domain as a coarse grained service: Business functions are driven by needs of the enterprise, rather than internal implementation details. Business functions may call finer-grained technical services within the domain internally, providing a service to external consumers. The main goal of defining business functions is limiting interactions between domains to well-defined points.
- Adding *gateways* that translate assumptions of one domain into assumptions of another domain: Gateways also allow establishing clear separation between external entities (business partners, for example) adding better control capabilities. Often, gateways may be combined with business functions into a single logical entity. However, adding a gateway is dictated by business requirements and may not be needed if domains share the same assumptions. Gateways could be implemented using one of the following approaches:
  - *Transparent/proxy gateways* expose domain services in such a way that a consumer believes that it interacts with a business function directly. Such gateways may also perform transformation of incoming and outgoing messages based on enterprise infrastructure requirements. Transformation may be required when domain data is stored in a very specific format that is radically different from enterprise-wide conventions. For example, domain-specific

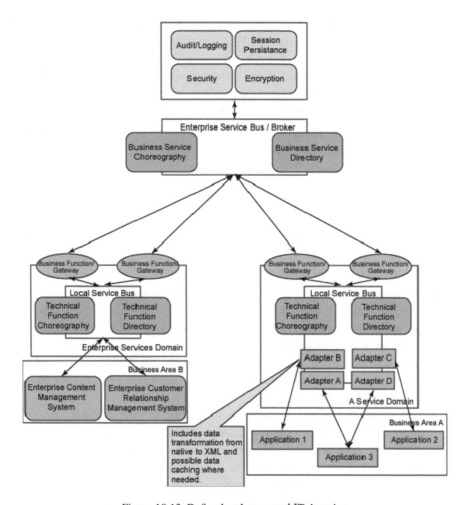

Figure 10.13 Defined and separated IT domains.

binary data may need to be transformed into an XML per enterprise policies.

– *Firewalls* allow for enforcement of the domain encapsulation. Although, they do not perform any transformation, firewalls limit consumer access to only predefined points in the domain boundaries. Also, introduction of a firewall allows detection of services that do not follow encapsulation policy of a domain.

When analyzing application functions to be exposed to the outside of a domain, one may find that the actual service is a combination of a finer grain functions. One mechanism to achieve such orchestration of application functions into a business services is to introduce a Local Service Bus (LSB), with the addition of a Service Choreography component into such domains.

Local Service Bus is an instance of an Enterprise Service Bus that provides connectivity support for a single domain. Being extended with the Service Choreography component allows the domain to compose higher level business functions that get consumed by the ESB.

Figure 10.13 demonstrates the infrastructure that may result from following the above steps.

Once all domains are encapsulated and business functions are exposed, it becomes easier to integrate them through the ESB and to employ Business Choreography to create higher level Business Processes and Transactions. Once the ESB is in place and the enterprise is functioning, the migration of legacy applications within domains may be performed with less impact on the rest of the enterprise.

10.4 Summary

We have presented the ESB and its role as the infrastructure underpinning an integrated and flexible SOA. Our presentation identified service metadata managed through a service registry as a key component of the ESB, allowing integration specialists and administrators to create and manage service-oriented solutions. Clear definition of the interfaces, and of the capabilities and requirements of the services, enables mediations to reconcile differences between service requestors and providers. We discussed a range of mediation patterns. We described ESB usage patterns in which these abstract concepts are applied to enterprise scenarios. These concepts are realized through a variety of technologies and products in a large and growing number of customer solutions, including large-scale retail and brokerage applications. ESB adoption and use is expected to continue at full strength for the foreseeable future, and the ESB plays a central role in the implementation of the architecture for the IBM on Demand Operating Environment.

Interest in EAI is driven by a number of important factors. With the pressures of a competitive business environment moving IT management to shorter application lifecycles, financial prudence demands that IT managers learn to use existing databases and application services rather than recreate the same business processes and data repositories over and over. EAI technology is maturing fatly and it has become a key topic for IT due to business imperatives. These include the emergence of the Web, the need to develop and deepen relationships with customers and partners, streamlining internal business processes, and more importantly reducing the time to market for new applications. In fulfilling these requirements, enterprises are highly motivated to make effective use of their existing custom-written legacy applications and commercial packaged applications, which are currently the enterprise's crown jewels. They are highly efficient and robust and above all there represent an enormous investment. But there are a lot of things to be understood before embarking on integration. They are what applications need integration, platforms, data formats, protocols, what EAI technology to be used, etc.

References

1. J. Schekkerman. *EA & Services Oriented Enterprise (SOE)*. EA Publications, http://www.roseindia.net/eai/enterpriseapplicationintegration.shtml, 2008.
2. M.T. Schmidt, B. Hutchison, P. Lambros, and R. Phippen. The Enterprise Service Bus: Making service-oriented architecture real. *IBM Systems Journal*, 44(4), 2005.
3. Web Services Architecture Overview. IBM DeveloperWorks. http://www.106.ibm.com/developerworks/web/library/w-ovr, September 2000.
4. J. Farrell, R. Akkiraju, and M.-T. Schmidt. Web Services Semantic Annotations, Technical Note, Version 1.0, http://awwebx04.alphaworks.ibm.com/ettk/demos/wstkdoc/services/demos/psme/webapp/WebServicesSemanticAnnotations.htm, 2004.

11

Commercialization

11.1 Overview of WSDC

A Web Services Distribution Channel is the collection of strategies, methods, technologies and relationships required to get an organization's Web services in front of customers in an appropriate form and at the most opportune time and place to dramatically increase usage.

This part of the definition shows that creating and maintaining a WSDC is much more complex than merely integrating disparate systems. No doubt the technologies involved are important. However, when services are exposed outside the organization for the purposes of creating new markets or expanding existing markets, more decisions need to be made with regard to the market (product, price, place, promotion) than technology. In fact, a well designed WSDC usually has a business plan backing it up that clearly defines the following:

- Product (service) and anticipated usage patterns.
- Financial model (including pricing matrix).
- Customer profile.
- Marketing plan.
- Competition (existing and future competitors).
- Success metrics.
- Risks (technical, brand and legal).

11.2 Creating WSDC

11.2.1 Process for Creating the Web Services Distribution Channel

Like most aspects of business, there is no one-size-fits-all strategy for achieving WSDC bliss. Since every organization is different in terms of goals, capabilities and markets served, no two WSDC implementations will look the

271

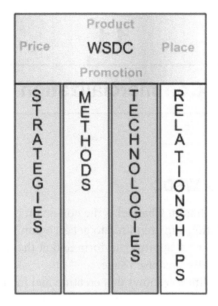

Figure 11.1 WSDC architecture.

same. However, there is a process within which an organization can engage to help ensure its WSDC is effective. At a high level, the keys to implementing an effective WSDC are:

1. Clearly understanding your organization's objectives (i.e. what business goal are you trying to achieve by exposing the proposed services?).
2. An appreciation for the Web Service Distribution Ecosystem (functions, roles and participants).
3. Knowing which questions to ask and how to interpret the answers.

Step 1 – Assess Service-Based Offerings

The creation of a WSDC requires a frank assessment of an organization's offering in terms of:

- market applicability;
- state of readiness;
- sustainability.

Market Applicability   Before entering a new market, it is critically important to know if customers want what you plan to sell and how they will use it. Moreover, you also need to know what alternatives will be available to your target customers. This being the case, the first part of assessing your service-based offering is determining its applicability in the marketplace by answering questions such as:

- Who is the target customer?
- What will their usage patterns look like?
- Who else is providing similar services?
- Who might enter the space if you are successful?
- What might customers pay for your services based on: the value the WSDC provides, their ability to pay, competing offerings and expectations?

State of Readiness Once you have determined there is a viable market for your services, you then need to perform a gap analysis to ascertain the steps required to make those services actually marketable. For example, are the services ready for delivery? If yes, are they in a consumable form? Are the required hosting and commerce components in place to facilitate the service as a business? Are there relationships upon which providing this service is critically dependent? In order to firm up the financial model that supports the WSDC, you need to intimately understand the effort and cost required to bring your services to your target market.

Sustainability Next, you need to accurately assess service maintenance requirements. During this process, you will need to factor in questions pertaining to operational, scalability and contingency issues. Remember, this channel 'is' an online business and must be highly available and reliable. Implementing a WSDC is not a fire-and-forget type of initiative. Finally, there is a growing number of service providers who can cost-effectively provide important pieces of support and maintenance infrastructure, so be sure to evaluate your options carefully.

Step 2 – Develop Financial Models and Marketing Plan

Now that you understand which services you will be selling, who will buy them, and what it is going to cost to deliver and maintain them, you are ready to construct financial models and develop a marketing plan. Depending upon the kinds of services you will be offering, creating financial models can be a complex process often requiring its own methodology.

The first step in creating a financial model is constructing a pricing matrix – a critically important component because the entire financial model stems from the revenue projections derived from anticipated pricing. Keep the following questions in mind when developing this part:

- What does it cost to produce these services?
- What do customers currently pay for similar services? What are their expectations?
- What are competitors charging?
- How will these services be assembled with other services that your organization or others produce?
- Do your services help customers increase revenues, reduce costs or merely facilitate existing transactions?
- How will the customer be charged for the service – pay per use, subscription, or included in a broader analog transaction?

Marketing Plan   Before you can build a full financial model, you need to understand how your services will be marketed and the associated costs of those marketing activities. This is where a number of organizations fail either because they grossly underestimate marketing costs or merely fail to see the need to market services at all. This happens often when external Web services initiatives are looked upon as technical systems integration projects and not as business channel activities. Remember, if the goal is to increase service usage and thus derive additional revenues, you need to make sure customers know about them, understand their value and how to use them.

After the marketing plan is created, you are ready to construct the financial model – tracking projected revenues to projected expenses. The completion of the financial model represents a seminal milestone because this model is the final piece of the puzzle required to make a 'go/no go' decision. If you have performed the proper due diligence and have confidence in your data, the decision to proceed or not should be coming into focus. While it is important to be passionate, be smart and honest in your appraisal. All the passion in the world is not going to make an unviable service successful.

Lawyers Support   While exposing Web services for public consumption can open many new doors of opportunity, it also brings with it a host of new legal concerns. If you have done your job right, it is quite possible your services will be consumed by parties and used in ways you did not anticipate. Moreover, it is even possible your services will be combined with services from other publishers to create new and unintended offerings. Because of this, you need to consider the legal costs and ramifications associated with creating the WSDC. Areas to consider include: service usage rights and responsibilities, the creation of derivative works, data retention policies and service level agreements.

Step 3 – Design

Now that your WSDC is backed by a sound strategic plan, it is time to get down into the technical weeds and design the actual services. You should stick with methodologies that are in tune with development team size and style, application development tooling, application runtime and hosting environment and security policies. However, there are a couple of issues to consider that transcend environmental factors:

Service Granularity   If the purpose of creating a WSDC is to increase usage of your service-based capabilities, then it is important to design your services at a level of granularity that supports your most common, and hopefully, profitable usage patterns. While there is no 'one right' way to organize candidate operations and services, there clearly are wrong ways, depending upon your objectives. As a rule of thumb, your services should be atomic, isolated and have clearly defined functional boundaries [1]. Remember that customers will often combine them with other services in ways you never anticipated so make sure the design is flexible enough to accommodate unforeseen use cases. Each new use represents a found revenue generation opportunity.

Leverage the Infrastructure   WSDCs assume a business grade runtime infrastructure that handles where needed: monitoring, metering, security, compression, optimization, scalability, contingency and commerce. This being the case does not build these functions into your services. Your services should merely contain the hooks required so the infrastructure can provide these support functions in a uniform and manageable manner to all WSDC services.

Step 4 – Implementation and Testing

If you have designed and documented your service architecture properly, developing, testing, and delivering the services should be a fairly straightforward, though not necessarily easy, process.

Step 5 – Investigate User Distribution Acceleration Options

While the Web services story to date has been mostly about internal systems integration projects, it has also been a 'programmer's story' in the sense that developers hold the key to the users' ability to access a service from within the applications they use. In other words, if a user wants an application to incorporate a function that exists in a Web service, a programmer must first modify the application's source code to embed the call, re-test and re-deploy the application. The cycle of service discovery, assessing its importance, modifying the application at the source code level, re-testing and re-deploying is the top inhibitor to the growth of Web services deployments. That means for every new WSDC you create, there may be a significant delay before customers incorporate your service into their applications.

The good thing is that there are technologies and techniques available that can help accelerate the delivery of your WSDC services into foreign applications. Think of them as packaging mechanisms that present your services when they are needed, where they are needed and in a form that is most applicable to the user. For example, there is a class of desktop integration software called indirect connectors that allow end users to integrate Web services with their applications without having to modify the application's source code or requiring the assistance of a programmer. Indirect connectors allow software users to link application screen fields to the inputs and outputs of Web services so data can be extracted from application screens, processed by Web services, and have the results pasted into the same or different application fields.

In addition, the tooling for composite software and mashups is getting more sophisticated every day so it is important to understand how these tools can help broaden the reach of your WSDC.

Step 6 – Execute Marketing Plan

The final step in implementing a WSDC is executing the marketing plan. In order for a WSDC to be successful, target customers need to be made aware of its existence and value. Awareness is created through a combination of innovative packaging techniques that are consistent with the way customers want to consume your services and partnerships with other providers who have a vested interest in your success.

This vested interest is created when: your WSDC creates additional value to the partner's offering, a partner's paid transaction is tied to your service, or the partner is compensated each time your service is executed. These marketing relationships come in all shapes and sizes and are more apt than any of your other channels to involve organizations with whom you have never considered partnering. So it is important to understand the Web Services Distribution Ecosystem and its participants in order to maximize the channel's value.

11.3 Web Service Distribution Eco System

The Web Services Distribution Ecosystem (WSDE) is the universe of roles and participants that facilitate the delivery of publicly available Web services to consumers. Since WSDCs are merely organization-specific overlays of these roles and participants, the key to creating an effective WSDC starts with an understanding of the WSDE. Figure 11.2 depicts the current view of the WSDE. As the concept matures, it is inevitable new roles will emerge while others will be either consolidated or deprecated.

Publisher
A publisher is a producer of unique content and/or functionality 'incentivized' to promote wide-scale adoption and reuse [2]. Publishers create or assemble the actual services consumers need to conduct business. Multiple publishers can participate in one WSDC or a publisher's offering can be part of multiple WSDCs.

Presence Provider
The presence provider is responsible for all the technical aspects of: hosting the service endpoint on the Internet, the service contract (in form of a WSDL), and all the plumbing required to connect to the publisher's services. The presence provider is also responsible for supplying the infrastructure and management functions required to make a service fault tolerant and scalable.

Payment Provider
The payment provider facilitates the mechanism used to pay for service access. These mechanisms may include credit card transactions, service hit counting or subscription tracking [3]. If a WSDC is based on a free service offering, the payment provider role will most likely be omitted.

Account Provider
The account provider authenticates the consumer's identity and authorizes service level access. It represents the primary holder of the user contact in-

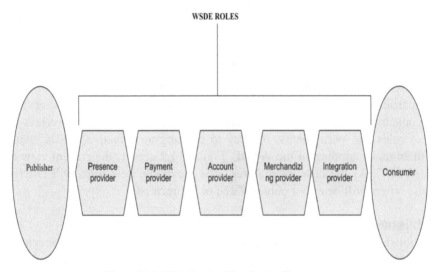

Figure 11.2 Web Service Distribution Ecosystem.

formation and is a role often consolidated within the merchandizing provider role described below.

Merchandizing Provider

The merchandizing provider promotes, prices and bundles services for sale to consumers. It serves as the WSDC's 'retail' outlet and, by doing so, often 'owns' the consumer relationship. This role represents the primary vehicle by which service awareness is created.

Integration Provider

The integration provider is responsible for integrating services into server-based, batch-based or desktop applications via direct integrations and/or indirect connectors.

Direct connections are made when an application, server-based process, BPM process, or mashup is modified at the source code level to contain explicit calls to a given service or proxy. While there are ways to more 'loosely couple' the connections between applications and specific services, source code level modifications must be made for an application to directly integrate with a WSDC. Direct connections often use an application's native API and usually facilitate deep integrations.

Indirect connections facilitate the integration of applications and WSDCs without requiring source code level changes to the application. These connections are established by external connectors that integrate with software from the 'outside-in' and do not require the knowledge or permission of the integrated application. These integrations tend to be 'more shallow' in scope with regard to any one specific application but allow end users to link a broad array of applications without coding or assistance from a developer.

Like merchandizing providers, integration providers play an important role in creating awareness for a given service offering.

11.4 SOA Trader Web Service Middleware

SOA Trader has been developing and fine-tuned its Web services commercialization middleware extension for simplifying delivery of services on-demand. While one of the development focuses has been separation of functional and non-functional specifications (i.e. QoS terms, licencing and subscriptions, etc.) of Web services, we have been considering also the aspects such as caching, QoS monitoring and error compensation. The result is a proxy-like light-weight servlet, which is easy to set-up and to deploy and provides good-enough performance in industrial settings.

The extension allows separating licencing and subscription options from Web service functionality such that service providers would be able to implement the main functionality of their Web services and all the rest, needed for supporting commercialization, would be provided by the extension. This means that you can shorten your technical service development cycles.

The extension acts as a proxy between Web service endpoints and the client application while providing effective SOAP cashing, service monitoring, service licensing and transparent error compensation. The middleware allows to define subscription packages for Web service clients, which are then used to facilitate subscription-based access to particular Web services, if needed [4]. One of the implemented robust and effective error compensation mechanism makes use of the caching functionality to provide results from data-provisioning services, even when the services' endpoints are inoperational. This can be switched on and off either at Web service or operation level similarly to caching mechanisms.

For each registered Web service an endpoint at proxy is constructed and made available for incoming requests. After a license has been acquired from the Web Services Marketplace, it can be used to access subscribed Web services. Service providers at the same time can define subscription packages to their services, which are then acquired by service requesters to access particular services. The QoS of brokered Web services is monitored constantly and published at SOA Trader's Web page along with detailed descriptions of particular services.

11.5 Summary

This chapter summarizes the functioning and working of Web Services Distribution Channel (WSDC) and how to create a WSDC component with its potentiality. A discussion on Web Service Distribution Eco-System and the architecture that supports the WSDE environment and concludes with the SOA architecture and its role as middleware.

References

1.   Thomas Erl. *Service-Oriented Architecture: Concepts, Technology, and Design*. Prentice Hall, 2005.
2.   Joe Labbe. Commercializing Services: Web Services Distribution Channels and SOA. *SOA Magazine*, III, 2007.
3.   Thomas Erl, *Service-Oriented Architecture: A Field Guide to Integrating XML and Web Services*. Prentice Hall/Pearson PTR.
4.   Sarah Perez. Distributed Channels for Mobile Applications. ReadWrite Mobile, 2010.

12

Emerging Standards and Development of EAI

12.1 Introduction

Enterprise Application Integration (EAI) is an integration framework composed of a collection of technologies and services which form a middleware to enable integration of systems and applications across the enterprise. Supply chain management applications, customer relationship management applications, business intelligence applications, and other types of applications typically cannot communicate with one another in order to share data or business rules. Such applications are sometimes referred to as islands of automation or information silos. This lack of communication leads to inefficiencies, wherein identical data are stored in multiple locations, or straightforward processes are unable to be automated.

Enterprise application integration (EAI) is the process of linking such applications within a single organization together in order to simplify and automate business processes to the greatest extent possible, while at the same time avoiding having to make sweeping changes to the existing applications or data structures. Generally, EAI is the 'unrestricted sharing of data and business processes among any connected application or data sources in the enterprise'.

One large challenge of EAI is that the various systems that need to be linked together often reside on different operating systems, use different database solutions and different computer languages, and in some cases are legacy systems that are no longer supported by the vendor who originally created them.

12.1.1 Purposes of EAI

EAI can be used for different purposes:

- Data (information) Integration: Ensuring that information in multiple systems is kept consistent. This is also known as EII (Enterprise Information Integration).
- Vendor independence: Extracting business policies or rules from applications and implementing them in the EAI system, so that even if one of the business applications is replaced with a different vendor's application, the business rules do not have to be re-implemented.
- Common Facade: An EAI system could front-end a cluster of applications, providing a single consistent access interface to these applications and shielding users from having to learn to interact with different software packages.

12.1.2 EAI Patterns

- *Integration patterns*: There are two patterns that EAI systems implement:
  - *Mediation*: Here, the EAI system acts as the go-between or broker between (interface or communicating) multiple applications. Whenever an interesting event occurs in an application , an integration module in the EAI system is notified. The module then propagates the changes to other relevant applications.
  - *Federation*: In this case, the EAI system acts as the overarching facade across multiple applications. All event calls from the 'outside world' to any of the applications are front-ended by the EAI system. The EAI system is configured to expose only the relevant information and interfaces of the underlying applications to the outside world, and performs all interactions with the underlying applications on behalf of the requester [1].

  Both patterns are often used concurrently. The same EAI system could be keeping multiple applications in sync (mediation), while servicing requests from external users against these applications (federation).
- *Access patterns*: EAI supports both asynchronous and synchronous access patterns, the former being typical in the mediation case and the latter in the federation case.
- *Lifetime patterns*: An integration operation could be short-lived e. g., keeping data in sync across two applications could be completed within a second or long-lived e. g., one of the steps could involve the EAI

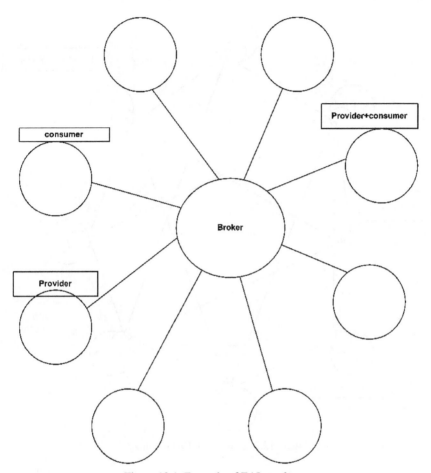

Figure 12.1 Example of EAI topology.

system interacting with a human work flow application for approval of a loan that takes hours or days to complete.

12.1.3 EAI Topologies

There are two major topologies: hub-and-spoke, and bus:

- *Hub topology*: In the hub-and-spoke model, the EAI system is at the center (the hub), and interacts with the applications via the spokes. In the bus model, the EAI system is the bus (or is implemented as a res-

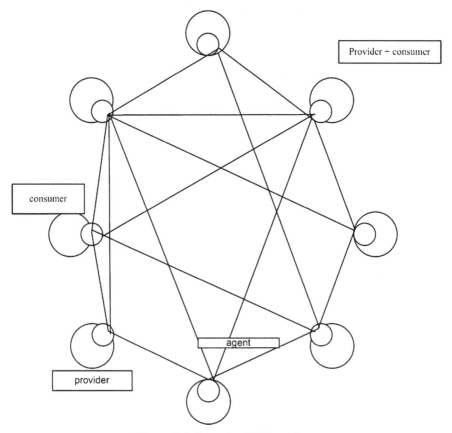

Figure 12.2 Example of EAI topology.

ident module in an already existing message bus or message-oriented middleware).

- *Bus topology*: Multiple technologies are used in implementing each of the components of the EAI system, for example bus/hub. This is usually implemented by enhancing standard middleware products (application server, message bus) or implemented as a stand-alone program (i.e., does not use any middleware), acting as its own middleware.

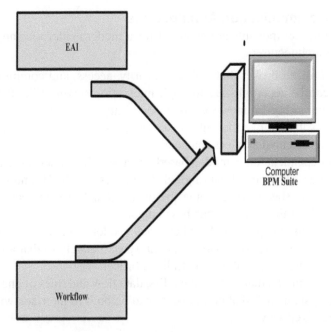

Figure 12.3 EAI and workflow.

12.1.4 EAI Technologies

12.1.4.1 Application Connectivity

The bus/hub connects to applications through a set of adapters (also referred to as connectors). These are programs that know how to interact with an underlying business application. The adapter performs two-way communication, performing requests from the hub against the application, and notifying the hub when an event of interest occurs in the application (a new record inserted, a transaction completed, etc.)

Adapters can be specific to an application (e.g., built against the application vendor's client libraries) or specific to a class of applications (e. g., can interact with any application through a standard communication protocol, such as SOAP or SMTP). The adapter could reside in the same process space as the bus/hub or execute in a remote location and interact with the hub/bus through industry standard protocols such as message queues, Web services, or even use a proprietary protocol. In the Java world, standards such as JCA allow adapters to be created in a vendor-neutral manner.

12.1.4.2 Communication Architectures

There are four components are essential for a modern enterprise application integration architecture:

1. A centralized broker that handles security, access, and communication. This can be accomplished through integration servers (like the School Interoperability Framework (SIF) Zone Integration Servers) or through similar software like the Enterprise service bus (ESB) model that acts as a SOAP-oriented services manager.
2. An independent data model based on a standard data structure, also known as a Canonical data model. It appears that XML and the use of XML style sheets has become the de facto and in some cases de jure standard for this uniform business language.
3. A connector, or agent model where each vendor, application, or interface can build a single component that can speak natively to that application and communicate with the centralized broker.
4. A system model that defines the APIs, data flow and rules of engagement to the system such that components can be built to interface with it in a standardized way.

Although other approaches like connecting at the database or user-interface level have been explored, they have not been found to scale or be able to adjust [2]. Individual applications can publish messages to the centralized broker and subscribe to receive certain messages from that broker. Each application only requires one connection to the broker. This central control approach can be extremely scalable and highly evolvable.

Enterprise Application Integration is related to middleware technologies such as message-oriented middleware (MOM), and data representation technologies such as XML. Other EAI technologies involve using Web services as part of SOA as a means of integration. Enterprise Application Integration tends to be data centric.

12.2 Java Enterprise Suites

The Java Enterprise System introduced a radical new approach to enterprise infrastructure software, the entire process of how infrastructure software is acquired, deployed, and operated was made simple, predictable, and affordable. And now with Java System Suites, that simple process just became easier. Java System Suites are designed to simplify the way customers use the Java Enterprise System to solve their business problems. The Java Enterprise System brings the best in Java Web and application infrastructure together to address targeted business needs in the financial services, government, telecommunications, manufacturing, and healthcare industries. It delivers enormous value to IT organizations by providing business agility, security, and optimization solutions that help reduce infrastructure software cost and complexity. With the Java Enterprise System, enterprises are saving millions of dollars in software license charges and operational costs, and dramatically simplifying how they select, buy, and use their infrastructure software assets. Java System Suites extend this value by enabling customers to choose from different solutions specific to the problems they want to solve.

As their business needs change and resources are freed up to tackle new projects, customers can adopt additional suites to address their needs incrementally. The Java Enterprise System comprises an integrated set of core enterprise infrastructure solutions. Designed and built with open industry standards, it is the smart choice for end-user customers, ISVs, OEMs, SIs, and resellers. To gain the benefits of the Java Enterprise System, customers can buy a subscription license for a suite or a combination of suites that includes the point products that address an immediate business need. Customers can also choose to license the complete Java Enterprise System including the full portfolio of Java Enterprise System products and pay a single annual license fee for the infrastructure software, support, maintenance, consulting, training, and education services. By adopting the Java Enterprise System model, customers can spend more time focusing on their business requirements rather than integrating and supporting a myriad of point products and can lay the foundation to move toward a more Service-Oriented Architecture (SOA). The result is better control and increased business agility for accelerated business service deployment.

12.2.1 Key Feature Highlights

The Java Enterprise System keeps it simple:

- *Simplicity* – Reduced complexity and cost to acquire, deploy, and operate infrastructure software.
- *Predictability* – Regular releases and terms to help reduce infrastructure software variables in licensing, planning, and deployment.
- *Affordability* – Lower cost to acquire and operate infrastructure software assets and related services Java System Suites provide subsets of the Java Enterprise System component products to extend the value of the Java Enterprise System. The suites offer:

– A pragmatic approach to solving business problems.

– An infrastructure foundation that enables customers to move towards a SOA.

– Multiple entry-points for getting started with enterprise infrastructure building blocks, using the following suites:

– Java Application Platform Suite.

– Java Identity Management Suite.

– Java Communications Suite.

– Java Availability Suite.

– Java Web Infrastructure Suite.

12.2.2 Java System Suites – Targeted at Solving Business Problems

Java System Suites introduce greater flexibility for how customers can leverage the benefits of the Java Enterprise System. The suites are designed to address customers' top-of-mind concerns, and provide a foundation to make their business:

- *Agile* – Responsive to market demands and competitive pressures
- *Secure and Regulatory Compliant* – Confident that with secure access and ID control, IT processes comply with ever changing rules and regulations
- *Optimized* – Able to expedite consolidation, rationalization, and integration across all services through an open, integrated, and standards-based IT infrastructure.

12.2.3 Business-Centric Suites

Java System Suites group together functionally similar Java Enterprise System product components so that customers can start implementing the Java Enterprise System. As their business requires new infrastructure services, customers can combine suites to fit their growing business needs. The integrated suites offer a new way to deliver specific groupings of Java Enterprise System product components to customers. They will provide better integration, greater predictability, increased flexibility, more reusable business processes, and higher security beyond the individual core components, and enable a phased-in approach to the overall Java Enterprise System over time.

The following suites are available:

- Java Application Platform Suite;

- Java Identity Management Suite;
- Java Communications Suite;
- Java Availability Suite;
- Java Web Infrastructure Suite.

12.2.4 Integrated Application Platform

The Java Application Platform Suite is a comprehensive, flexible, secure, and reliable platform designed to accelerate time to service of new applications and streamline integration of legacy applications through a robust portal platform. Bringing together the market-leading Application, Web, and Portal services of the Java Enterprise System with a comprehensive set of developer tools, the Application Platform Suite is a pragmatic first step in implementing a SOA. The suite combines and maximizes the benefits of the standard Java 2 Platform, Enterprise Edition.

J2EE architecture, fostering application reuse with the highly scalable and secure Java System Web Server. The Application Platform Suite enables fast and accurate development and delivery which is an essential competency to virtually all companies today through a complete set of Java developer tools that support the latest standards, dynamic developer collaboration, and SOA methodologies. Secure universal access is critical to ensuring high usage rates of new applications and services. The Application Platform Suite enables access from outside the company firewall securely from virtually any Web browser and supports access from over 4000 mobile devices. In addition to delivery to remote and mobile users, applications and services can be readily delivered based on users' roles and privileges. The Java System Portal Server provides a robust delivery platform that includes powerful personalization, aggregation, integration, and search capabilities.

12.2.5 Datasheet Java Enterprise System and Java System Suites

The Java Enterprise System offers a single, comprehensive software system with all of the critical enterprise infrastructure elements you need, packaged with end-to-end support, maintenance, consulting, training, and education – all delivered at regular, predictable intervals.

The Application Platform Suite excels against the competition by providing unmatched value in the most comprehensive platform for IT organizations worldwide, and includes the following world-class infrastructure products:

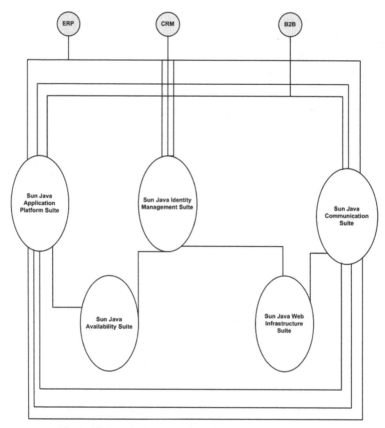

Figure 12.4 Infrastructure DNA for business applications.

- Java System Application Server Enterprise Edition;
- Java System Web Server;
- Java System Portal Server;
- Java System Portal Server Secure Remote Access;
- Java System Portal Server Mobile Access;
- Java Studio Enterprise;
- Java Studio Creator.

12.2.6 End-to-End Identity Management

The *Java Identity Management Suite* offers the most comprehensive, innovative suite of identity management solutions designed to secure, streamline, and simplify the process of managing user identities across globally dispersed communities, computing infrastructures, and application environments. The full suite of solutions eliminates costly manual creation, maintenance, and deletion of identity data, enabling organizations to increase accessibility while maintaining tight security. To help ensure compliance with regulatory requirements identity management offerings provide centralized control, complete visibility into access privileges, and consistent enforcement of identity management policies across the enterprise. Identity Management Suite provides the core functions and services required by enterprises to use, share, and manage identity information, and include:

- Java System Identity Manager: The industry's first product to converge user provisioning and metadirectory capabilities for efficiently and securely managing identity profiles and permissions throughout the entire identity lifecycle.
- Java System Access Manager: The industry's first commercially available access management solution providing open, standards-based access control, single sign-on, and federation services to comply with the Liberty Phase 2 and SAML 1.1 specifications.
- Java System Directory Server Enterprise Edition: The industry's first directory services solution to deliver enterprise-level services, including built-in failover, load balancing, and security and integration with Microsoft Active Directory. It provides a secure, highly available, scalable, and easy-to-manage directory infrastructure that effectively manages identities in growing and dynamic environments.

12.2.7 Secure, Feature-Rich Communications

The *Java Communications Suite* enables secure, reliable delivery of a rich set of communication and collaboration services messaging, real-time collaboration, calendaring, and scheduling at less than half the cost of alternative solutions. Today's enterprises can improve employee productivity, customer satisfaction, and partner relationships by expanding communication and collaboration services to broader user constituencies and, at the same time, lower the cost of providing those services. Service providers looking to attract new customers including enterprises and retain existing customers with expanded and differentiated services can lower their bottom line with the Communications Suite. High performance and scalability enable efficient communications [3].

Reliability and availability help ensure continuous service and multiple device support and secure remote access enable anytime, anywhere access. In addition to the broad communication and collaboration feature set, the Communications Suite delivers extensive security features such as user authentication, message and session encryption, and appropriate content filtering to help prevent spam and viruses. Identity-based access policy, user management, and end-user privacy controls also help ensure the integrity of communication services and facilitate compliance with industry regulations. Support for open Internet standards, well documented programming interfaces, and a modular architecture help protect communications infrastructure investments making it easier for Java Enterprise System customers to meet the changing needs of their dynamic business environments. Components of the suite also provide building blocks for enhanced communications and presence-enabled applications.

With the Communications Suite, enterprises and service providers can build the secure, reliable communication and collaboration services required to meet their business needs. The Communications Suite includes the following industry-leading and award-winning products:

- Java System Messaging Server;
- Java System Instant Messaging;
- Java System Calendar Server;
- Java System Connector for Microsoft Outlook;
- Java System Synchronization Tool;
- Java Studio Enterprise;
- Java Studio Creator.

12.2.8 Mission-Critical Availability

The *Java Availability Suite* offers a foundation for delivering higher levels of availability at lower costs. Leveraging the predictable Java Enterprise System subscription and regular release model, near-continuous application availability can be achieved with tighter integration and reduced risk. High-availability (HA) applications deployed on the Availability Suite leverage the robust framework that is tested and designed to work with other components of the Java Enterprise System. This suite enables easier management of the overall HA environment. The Availability Suite is an integrated solution that offers:

- Service-level management of mission-critical applications.
- Industry-leading data and application high availability.
- Powerful and easy-to-use manageability capabilities.

The Availability Suite enables customers to harness the industrial-strength high availability of Cluster software with the agility, security, and integration benefits of Java System Suites. It includes:

- Cluster;
- Cluster Agents;
- Java Studio Enterprise;
- Java Studio Creator.

12.2.9 Secure, Reliable Web Services

The *Java Web Infrastructure Suite* provides an affordable, quick-start approach to Web services, with the simplicity, ease of use, and out-of-the-box integration of a more fully featured platform. Designed as a secure, reliable platform for enterprise workgroup and departmental applications, the Web Infrastructure Suite also addresses mid market solutions.

Web services created with the Web Infrastructure Suite help organizations align their IT architecture with real-time business requirements, giving them the flexibility to add new services in response to market dynamics and competitive offerings. In addition, by exposing common business functions as services, tremendous gains in productivity and efficiency can be achieved through service reuse and centralization.

With industry-leading security and performance, developers and system administrators can be confident that their Web services will be always available and protected from intruders. The Web Infrastructure Suite includes all the products and tools customers need to securely and reliably develop and deploy Web services. The suite is available for subscription or with perpetual license pricing, and includes the best-of-breed Web services platform products:

- Java System Application Server Standard Edition;
- Java System Web Server;
- Java System Web Proxy Server;
- Java System Directory Server (limited license);
- Java System Access Manager (limited license);
- Java Studio Enterprise;
- Java Studio Creator.

12.2.10 Java Enterprise System

The Java Enterprise System introduced three radical and industry-changing elements with its first edition in January 2004:

1. A new software system of open, industry leading enterprise infrastructure solutions integrated as one entity.
2. A new aystematic approach to designing, deploying, and operating the software system.
3. A new business model that provides a single price and a single license for the software system, maintenance, support, consulting, training, and education services. These elements are core to the Java Enterprise System strategy and design.

12.2.11 A Core Set of Integrated, Enterprise

12.2.11.1 Infrastructure Services
Enterprise infrastructure services are the capabilities that sit between the traditional operating system such as the Solaris OS or Linux OS and the business applications. Enterprise infrastructure services are engineered and deployed to meet the business requirements for a scalable, interoperable, available, and secure IT infrastructure.

12.2.11.2 Common Features and Standards
The Java Enterprise System and Java System Suites deliver a software system of shared components, common technologies, a consistent architecture, and consistent user experience. The following are just a few of the key highlights of the software system:

- A unified infrastructure, providing data consistency, secure user identity, and easy addition of new identity-enabled applications.
- Single sign-on for all services, delivering a superior user experience and dramatically reducing help desk costs.
- Common installation, simplifying and improving deployment times and maintenance.
- Consistent support for high-availability deployment of system components.
- Interoperability features for ease of operation in established environments, including identity management and portal support for IBM, and BEA application servers; plug-and-play support for standards-based

portlets and support for Outlook and Evolution messaging and calendar clients for maximum efficiency without the cost.

- Solaris 10 support for leading security, management, and performance.
- Broader availability on high-performance AMD Opteron processor-based systems.
- Integrated Java System Application Server Enterprise Edition (J2EE 1.4) for improved integration,
- Java 2 Platform, Standard Edition (J2SE^{TM} platform) 5.0 support for added Java scalability and performance.

12.2.12 Consistent User Experience

12.2.12.1 Alignment of User Experiences

The Java Enterprise System provides a system wide, uniform user experience for continuity across all aspects of a user's interaction. Whether for business, consumers, or administrators, user interfaces adhere to a core set of standards to provide a seamless user experience. Common aspects of the user experience include:

- A core set of documents across the software system for all component products.
- Graphical user interface, help content, and key documents which all translated into eight different languages.
- Support for accessibility requirements across the entire software system.
- Adherence to usability style sheets and guidelines across all user interfaces, including command line interfaces, help display, graphical user interfaces, and others.

12.2.13 Reference Architectures

The Java Enterprise System is tested against common customer usage patterns and deployment practices, and is set up in environments that simulate actual deployments. Software system testing uses real deployment patterns that provide a virtual turnkey setup for deployments. Some of the information available to set up and optimize a Java Enterprise System deployment includes:

- Tuning and sizing information.
- Deployment best practice guides.
- Architecture recommendations for varying customer needs.
- Performance benchmarks for varying sized deployments.

- Reference architectures.

12.2.14 Support and Consulting Services Included

Consulting, proactive support, and in-depth education services help customers architect, implement, and manage their Java Enterprise System environment. Services offer comprehensive architecture, implementation, and management services and methodologies that enable customers to take full advantage of the Java Enterprise System. Services provides customers with technical support, software maintenance, installation services, custom consulting, and comprehensive education to ensure smooth transition and integration of the Java Enterprise System into their enterprise. Enterprise customers receive varying levels of services as part of their Java Enterprise System purchase. The number of employees licensed for the Java Enterprise System dictates the level of service included.

12.2.15 Software Support Services

At a minimum, the Java Enterprise System license includes ongoing technical support and complete software maintenance with full access to updates and upgrades. These services are part of the Software Standard Support Services offering.

12.3 Technologies Using Java WSDP 1.5

12.3.1 Web Services-Based Technologies and Tools

Java WSDP is a package that provides early access to the latest technology implementations and tools for Web services development. The most current version of the package, Java WSDP 1.5, includes implementations of a number of Web services technologies that are not in J2EE 1.4, as well as newer releases of the Web services technology implementations that are in the J2EE 1.4 SDK. The implementations of Web services technologies that are not in J2EE 1.4 are: Java Architecture for XML Binding (JAXB) v1.0.4, lXML and Web Services Security v1.0, Service-Oriented Architecture and Web Services: Concepts, Technologies, and Tools XML Digital Signatures v1.0 EA2. The newer releases of Web services technology implementations that are in the J2EE 1.4 SDK are: Java API for XML Processing (JAXP) v1.2.6_01, Java API for XML-based RPC (JAX-RPC) v1.1.2_01, SOAP with Attachments API for Java (SAAJ) v1.2.1_01, Java API for XML Registries

(JAXR) v1.0.7. In addition to these Web services technology implementations, Java WSDP 1.5 also provides an Early Access implementation of the Java Streaming XML Parser (SJSXP), which implements the Streaming API for XML (StAX).

12.4 WS-BPEL

Business Process Execution Language (BPEL), short for *Web Services Business Process Execution Language* (WS-BPEL) is an OASIS standard executable language for specifying actions within Business processes with Web services.

Language export and import information by using Web Service interfaces exclusively.

Web service interactions can be described in two ways: executable business processes and abstract business processes. Executable business processes model actual behavior of a participant in a business interaction. Abstract business processes are partially specified processes that are not intended to be executed. An Abstract Process may hide some of the required concrete operational details. Abstract Processes serve a descriptive role, with more than one possible use case, including observable behavior and/or process template. WS-BPEL is meant to be used to model the behavior of both Executable and Abstract Processes.

WS-BPEL provides a language for the specification of Executable and Abstract business processes. By doing so, it extends the Web Services interaction model and enables it to support business transactions. WS-BPEL defines an interoperable integration model that should facilitate the expansion of automated process integration both within and between businesses.

The origins of BPEL can be traced to WSFL and XLANG. It is serialized in XML and aims to enable programming in the large. The concepts of *programming in the large* and *programming in the small* distinguish between two aspects of writing the type of long-running asynchronous processes that one typically sees in business processes.

Programming in the large generally refers to the high-level state transition interactions of a process. BPEL refers to this concept as an Abstract Process. A BPEL Abstract Process represents a set of publicly observable behaviors in a standardized fashion [4]. An Abstract Process includes information such as when to wait for messages, when to send messages, when to compensate for failed transactions, etc. *Programming in the small*, in contrast, deals with short-lived programmatic behavior, often executed as a single transaction and involving access to local logic and resources such as files, databases, etc. BPEL's development came out of the notion that programming in the large and programming in the small required different types of languages.

12.4.1 History

IBM and Microsoft had each defined their own, fairly similar, 'programming in the large' languages, WSFL and XLANG, respectively. With the popularity and advent of BPML, and the growing success of BPMI.org and the open BPMS movement led by JBoss and Intalio Inc., IBM and Microsoft decided to combine these languages into a new language, BPEL4WS. In April 2003, BEA Systems, IBM, Microsoft, SAP and Siebel Systems submitted BPEL4WS 1.1 to OASIS for standardization via the Web Services BPEL Technical Committee. Although BPEL4WS appeared as both a 1.0 and 1.1 version, the OASIS WS-BPEL technical committee voted on 14 September 2004 to name their spec WS-BPEL 2.0. This change in name was done to align BPEL with other Web Service standard naming conventions which start with WS – and accounts for the significant enhancements between BPEL4WS 1.1 and WS-BPEL 2.0. If not discussing a specific version, the moniker BPEL is commonly used.

In June 2007, Active Endpoints, Adobe Systems, BEA, IBM, Oracle and SAP published the BPEL4People and WS-HumanTask specifications, which des cribe how human interaction in BPEL processes can be implemented.

12.4.2 BPEL Design Goals

There were ten original design goals associated with BPEL:

1. Define business processes that interact with external entities through Web Service operations defined using WSDL 1.1, and that manifest themselves as Web services defined using WSDL 1.1. The interactions are 'abstract' in the sense that the dependence is on portType definitions, not on port definitions.
2. Define business processes using an XML-based language. Do not define a graphical representation of processes or provide any particular design methodology for processes.
3. Define a set of Web service orchestration concepts that are meant to be used by both the external (abstract) and internal (executable) views of a business process. Such a business process defines the behavior of a single autonomous entity, typically operating in interaction with other

similar peer entities. It is recognized that each usage pattern (i.e. abstract view and executable view) will require a few specialized extensions, but these extensions are to be kept to a minimum and tested against requirements such as import/export and conformance checking that link the two usage patterns.

4. Provide both hierarchical and graph-like control regimes, and allow their use to be blended as seamlessly as possible. This should reduce the fragmentation of the process modeling space.

5. Provide data manipulation functions for the simple manipulation of data needed to define process data and control flow.

6. Support an identification mechanism for process instances that allows the definition of instance identifiers at the application message level. Instance identifiers should be defined by partners and may change.

7. Support the implicit creation and termination of process instances as the basic lifecycle mechanism. Advanced lifecycle operations such as 'suspend' and 'resume' may be added in future releases for enhanced lifecycle management.

8. Define a long-running transaction model that is based on proven techniques like compensation actions and scoping to support failure recovery for parts of long-running business processes.

9. Use Web Services as the model for process decomposition and assembly.

10. Build on Web services standards (approved and proposed) as much as possible in a composable and modular manner.

BPEL's focus on modern business processes, plus the histories of WSFL and XLANG, led BPEL to adopt Web services as its external communication mechanism [3]. Thus BPEL's messaging facilities depend on the use of the Web Services Description Language (WSDL) 1.1 to describe outgoing and incoming messages.

In addition to providing facilities to enable sending and receiving messages, the BPEL programming language also supports:

- A property-based message correlation mechanism.
- XML and WSDL typed variables.
- An extensible language plug-in model to allow writing expressions and queries in multiple languages: BPEL supports XPath 1.0 by default.
- Structured-programming constructs including if-then-elseif-else, while, sequence (to enable executing commands in order) and flow (to enable executing commands in parallel).

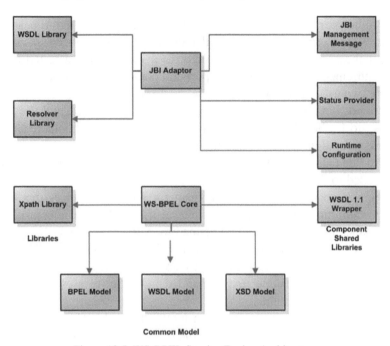

Figure 12.5 WS-BPEL Service Engine Architecture.

- A scoping system to allow the encapsulation of logic with local variables, fault-handlers, compensation-handlers and event-handlers.
- Serialized scopes to control concurrent access to variables.

12.4.3 Service Architecture

The BPEL Serice Engine supports business processes that conform to Web Services Business Process Execution Language (WS-BPEL) 2.0 specification. It provisions and consumes Web services described in WSDL1.1. It exchanges messages in JBI-defined XML document format for wrapped WSDL 1.1 message parts. The BPEL Service Engine can be configured in one of the following three modes: static, deployment, or runtime:

- *Static*: Parameter values, once loaded, can only be modified by re-installing the engine.
- *Deployment*: Parameter values can be changed without re-installation, but only until the engine is started/restarted; they remain in effect throughout business process execution.

- *Runtime*: Parameter values can be changed even while business processes are running.

The BPEL Service Engine supports request/reply, asynchronous one-way invokes, and direct invocation between two business processes. It supports monitoring of endpoint status. In addition, it also offers a command line facility to build the service assembly and test the deployed service.

12.4.4 WS-BPEL Engine Runtime Configuration

To support component configuration at installation and run-time our components follow the following conventions not covered by the JBI spec:

- Component configuration defaults are present in the JBI descriptor of the component.
- These defaults can be overridden via JBI installation configuration parameters at installation time. This is how our Web GUI allows these to be changed.
- At run-time, a custom JMX MBean exposes the configuration and allows relevant settings to be changed on-the-fly.

12.4.5 Advantages and Disadvantages of EAI Advantages

- Real time information access among systems.
- Streamlines business processes and helps raise organizational efficiency.
- Maintains information integrity across multiple systems.
- Ease of development and maintenance.

12.4.6 Disadvantages

- High initial development costs, especially for small and mid-sized businesses (SMBs).
- Require a fair amount of up front business design, which many managers are not able to envision or not willing to invest in. Most EAI projects usually start off as point-to-point efforts, very soon becoming unmanageable as the number of applications increase.

12.4.7 The Future of EAI

EAI technologies are still being developed and there still is no consensus on the ideal approach or the correct group of technologies a company should

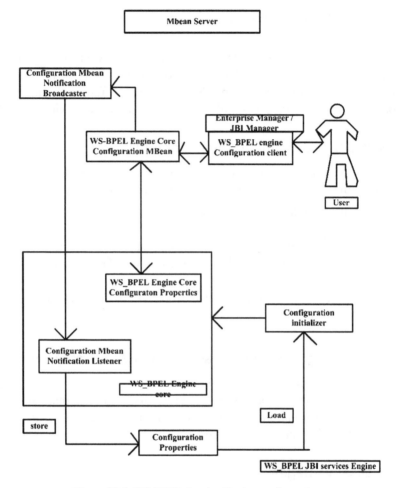

Figure 12.6 WS-BPEL Service Engine configuration.

use [2]. There are many ongoing projects that provide support to design EAI solutions. The future is to provide technologies that allow the design of EAI solutions at a high level of abstraction and use MDA to automatically transform the design into an executable solution. A common pitfall is to use other proprietary technologies that claim to be open and extensible but create vendor lock-in.

12.5 Summary

This chapter is a summary of the emerging standards and trends in the integrated application of SOA and Web Services. It also highlights on the support provided by the Java technical suite to build an integrated system approach. A detailed description of WS-BPEL (Business Process Execution Language) is presented and the chapter concludes with a forecast on the future of EAI technologies.

References

1. Gable and Julie. AIIM International, Enterprise Applications: Adoption of E-Business and Document Technologies, 2000–2001: Worldwide Industry Study. *Information Management Journal*, http://findarticles.com/p/articles/mi_qa3937/is_200203/ai_n9019202, 2008.
2. Thilina Gunarathne, Chathura Herath, Eran Chinthaka, and Suresh Marru. Experience with Adapting a WS-BPEL Runtime for eScience Workflows. Pervasive Technology Institute Indiana University, Bloomington.
3. Frank Leymann, Dieter Roller, and Satish. Design Goals of the BPEL4WS Specification. OASIS WS-BPEL Technical Committee, 2007.
4. Sun Java Enterprise System Deployment Planning White Paper, Sun Microsystems, Inc. 4150 Network Circle Santa Clara, U.S.A.

Index

Authors' Biographies

N. Sudha Bhuvaneswari has completed her MCA and MPhil in Computer Science and is currently pursuing her doctorate in Computer Science. Her area of interest is Mobile Agent Technology. She has been participating continuously in research and development activities for the past six years. She has presented and published technical papers in International Journals, at International Conferences and International Workshops organized by various international bodies like IEEE, WSEAS, and IEEE Explore. She has also contributed chapters in books like *Personal Area Network* and published articles and working manuals in agent technology. The author is currently employed as Associate Professor at the Dr. G.R Damodaran College of Science, Coimbatore, India. She is an active member of various technical bodies like ECMA, Internet Society of Kolkata and Chennai and acts as a moderator in various international conferences and journals.

S. Sujatha started as a Computer Science graduate. She completed her undergraduate degree at Sri Sarada College for Women, Tirunelveli and has also completed post graduate level courses MCA and MPhil at Bharathiar University, Coimbatore, India, and is currently pursuing her doctorate in Computer Science. Her of interest is Mobile Agent Technology & Networks. She has been participating continuously in research and development activities for the past six years. To her credit, she has presented and published technical papers in International Journals, at International Conferences and International Workshops organized by various international bodies like IEEE, WSEAS, and IEEE Explore. She has also contributed chapters in books like *Personal Area Network* and published articles and working manuals in agent technology. The author is currently employed as Associate Professor at the Dr. G.R Damodaran College of Science, Coimbatore, India. She is an active member of various technical bodies like ECMA, Internet Society of Kolkata and Chennai and acts as a moderator in various international conferences and journals.